T0260724

Performability Modelling:
Techniques and Tools

Performability Modelling:

Techniques and Tools

Edited by

Boudewijn R. Haverkort
RWTH Aachen, Germany

Raymond Marie, Gerardo Rubino
IRISA, Rennes, France

Kishor Trivedi
Duke University, USA

JOHN WILEY & SONS, LTD

Chichester • New York • Weinheim • Brisbane • Singapore • Toronto

Copyright © 2001

John Wiley & Sons Ltd
Baffins Lane, Chichester,
West Sussex, PO19 1UD, England
National   01243 779777
International (+44) 1243 779777

e-mail (for orders and customer service enquiries): cs-books@wiley.co.uk

Visit our Home Page on http://www.wiley.co.uk or http://www.wiley.com

All Rights Reserved. No part of this publication may be reproduced, stored in a retrieval system, or transmitted, in any form or by any means, electronic, mechanical, photocopying, recording, scanning or otherwise, except under the terms of the Copyright, Designs and Patents Act 1988 or under the terms of a licence issued by the Copyright Licensing Agency, 90 Tottenham Court Road, London W1P 9HE, UK, without the permission in writing of the Publisher and the copyright owner, with the exception of any material supplied specifically for the purpose of being entered and executed on a computer system, for the exclusive use by the purchaser of the publication.

Neither the author(s) nor John Wiley & Sons, Ltd accept any responsibility or liability for loss or damage occasioned to any person or property through using the material, instructions, methods or ideas contained herein, or acting or refraining from acting as a result of such use. The author(s) and Publisher expressly disclaim all implied warranties, including merchantability of fitness for any particular purpose.

Designations used by companies to distinguish their products are often claimed as trademarks. In all instances where John Wiley & Sons, Ltd is aware of a claim, the product names appear in initial capital or all capital letters. Readers, however, should contact the appropriate companies for more complete information regarding trademarks and registration.

Other Wiley Editorial Offices

John Wiley & Sons, Inc., 605 Third Avenue,
New York, NY 10158-0012, USA

Wiley-VCH Verlag GmbH
Pappelallee 3, D-69469 Weinheim, Germany

John Wiley & Sons Australia, Ltd, 33 Park Road, Milton,
Queensland 4064, Australia

John Wiley & Sons (Asia) Pte Ltd, 2 Clementi Loop #02-01,
Jin Xing Distripark, Singapore 129809

John Wiley & Sons (Canada) Ltd, 22 Worcester Road,
Rexdale, Ontario, M9W 1L1, Canada

Library of Congress Cataloging-in-Publication Data

Performability modelling: techniques and tools / edited by Boudewijn R. Haverkort (et al.).
       p. cm.
       Includes bibliographical references.
       ISBN 0-471-49195-0
       1.   Computer Systems—Evaluation. I. Haverkort, Boudewijn R.

       QA76.9.E94 P42 2001
       004.2'4—dc21
                                           2001017600

British Library Cataloguing in Publication Data
A catalogue record for this book is available from the British Library

ISBN 0-471-49195-0

Produced from PostScript files supplied by the author.

Printed and bound in Great Britain by CPI Antony Rowe, Eastbourne

Contributing authors

Andrea Bobbio University del Piemonte Orientale, Alessandria, Italy.
bobbio@di.unito.it

Nico M. van Dijk University of Amsterdam, Amsterdam, Netherlands.
nivd@fee.uva.nl

Lorenzo Donatiello University of Bologna, Bologna, Italy.
donat@cs.unibo.it

Richard Gail IBM Research, Yorktown Heights, USA.
rgail@research.ibm.com

V. Gopalakrishna Jataayu Software, Bangalore, India.
vgopi@jataayusoft.com

Vincenzo Grassi University of Rome "Tor Vergata", Rome, Italy.
grassi@info.roma2.it

Boudewijn R. Haverkort RWTH Aachen, Aachen, Germany.
haverkort@cs.rwth-aachen.de

Ranga Mallubhatla Bell Labs, Lucent Technolgies, NJ, USA.
rmallubhatla@lucent.com

Raymond Marie IRISA, Rennes, France.
marie@irisa.fr

John F. Meyer University of Michigan, Ann Arbor, MI, USA.
jfm@eecs.umich.edu

Isi Mitrani University of Newcastle, Newcastle, UK.
isi.mitrani@ncl.ac.uk

Krishna R. Pattipati University of Connecticut, Storrs, CT, USA.
krishna@engr.uconn.edu

Gerardo Rubino IRISA, Rennes, France.
`rubino@irisa.fr`

William S. Sanders University of Illinois, Urbana-Champaign, IL, USA.
`whs@crhc.uiuc.edu`

Perwez Shahabuddin Columbia University, New York, USA.
`perwez@ieor.columbia.edu`

Edmundo de Souza e Silva Federal University of Rio di Janeiro, Brazil.
`edmundo@nce.ufrj.br`

Miklós Telek Technical University of Budapest, Budapest, Hungary.
`telek@hit.bmu.hu`

Kishor S. Trivedi Duke University, Durham, NC, USA.
`kst@egr.duke.edu`

N. Viswanatham National University of Singapore, Singapore.
`mpenv@nus.edu.sg`

Contents

Foreword

THE concept of *performability* arose from the need to model and evaluate computer systems that exhibit *degradable performance*. More than two decades have passed since methods for this purpose were initially conceived and applied. During the same period, the fields of both computing and communication have experienced unprecedented growth, resulting in contemporary information systems that typically rely on a fusion of theories, techniques, and technologies from both fields. Consequently, it is not surprising that means of modelling and evaluating such systems have likewise evolved as an integration of concepts and methods originating from each discipline.

Generally, when modelling a system for evaluation purposes, one seeks to relate and quantify aspects of what the system is and does with respect to what the system is required to be and do. Moreover, since what a system does (e.g., how well it performs) depends on what it is (e.g., how its resources are altered by faults), both need to be addressed in the evaluation process. Just what is evaluated or, more precisely, the types of measures employed, can be classified according to certain assumptions regarding "is" and "does". In the context of computer/communication systems and with respect to a specified user-oriented or system-oriented service, *performance* typically refers to "quality of service, provided the system is correct". *Dependability*, according to current use of this term, is that property of a system which allows "reliance to be justifiably placed on the service it delivers". Such service is *proper* if it is delivered as specified; otherwise it is *improper*. System *failure* is identified with a transition from proper service to improper service. Dependability thus includes attributes of reliability and availability as special cases. Specifically, a reliability measure quantifies the "continuous delivery of proper service"; an availability measure quantifies the "alternation between deliveries of proper and improper service".

This basic distinction between performance and dependability has been particularly useful in the development of evaluation techniques suited to each concept. However, if separate evaluations of system performance and system dependability are to suffice in determining overall "quality of service" (e.g., QoS, as this term is defined and abbreviated by the telecommunication industry), one must place certain constraints on how properties affecting performance interact with those affecting dependability. For example, suppose that a system's capacity to serve is binary (either "up" or "down") and proper service coincides with that delivered when the system is up. In this case, performance refers to the quality of proper service and dependability measures the system's ability to remain up (and thus deliver that service) in the presence of faults. Accordingly, results of each type of evaluation, when taken together, can provide a rather complete assessment of overall service quality.

However, for information systems of even moderate complexity, individual assessments of performance and dependability are typically not combinable in this fashion. This is due to a phenomenon noted at the outset, namely that performance in the presence of faults is often degradable, i.e., fault-caused errors can reduce the quality of a delivered service even though that service, according to its specification, remains proper (failure-free). What is called for instead are performability measures which probabilistically quantify a system's "ability to perform" in a given operational environment. A specific performability measure is obtained by defining just what "performance" means for the evaluation in question. Choices here are virtually limitless, ranging from the type of binary-valued performance (success or failure) considered in dependability evaluation to continuous-valued performances such as throughput rates and processing delays. More generally, performance, when interpreted in a performability context, can be identified with *quality of service*, i.e., for a specified system service, its values reflect different degrees of satisfaction that might be experienced by a user of that service. With this interpretation, the corresponding performability measure quantifies the ability of a system to satisfy its users.

As attested to by the lengthy bibliography at the end of this monograph, the past 20+ years have witnessed an extensive amount of work concerning all aspects of model-based performability evaluation. This includes measure formulation, model specification, model construction, model solution, tool development, and application to a wide variety of systems. For the most part, progress in these areas has kept pace with the rapid growth in complexity of information systems and their environments. In particular, every effort has been made to continually expand the body of performability modelling theory and tools so as to enable the realization of increasingly ambitious evaluation studies. The various topics discussed in this monograph are representative of this effort, providing a sampling of some important enabling contributions. Without such results, which include the many tools surveyed in the final chapter, practical means of performability modelling and evaluation would not enjoy their current state of maturity.

John F. Meyer, Ann Arbor, Michigan
April 2001

Preface

IT has been a long time since the first idea of putting this book together was conceived. It was at the time of the 1993 performability workshop that the four co-editors of this book decided to pursue this idea.

The history of performability dates back to the late 1970s. Since then, many original results have been developed and published in various journals and conference proceedings. In the early 1990s the idea was then developed of having a small specialized workshop, purely devoted to performability modelling and evaluation. The first one was hosted by Haverkort and Niemegeers in 1991 at the University of Twente in the Netherlands. Subsequent workshops were held in Rennes, France (1993), in Chicago, Illinois (1996), and in the College of William and Mary in Williamsburg (1997). The fifth workshop will take place in Erlangen, Germany, in 2001.

From the workshops, only informal proceedings have been published, containing extended abstracts and work in progress reports. The first workshop had a special issue of the journal *Performance Evaluation* devoted to it (Volume 18, Issues 3 and 4). At the second workshop, the idea developed of compiling a book with specialist chapters, using unified notation, style, terminology and bibliographic references. Since then, a lot has happened in this challenging area, and many more results have been achieved, as represented by the current volume. We hope that this book will become a unique source on the topic of performability evaluation.

We would like to thank all the authors of the chapters for their concise writing, their patience, their hard work and their cooperation. Thanks are also due to Sarah Hinton of John Wiley & Sons for her help, patience and cooperation.

Boudewijn Haverkort, Raymond Marie, Gerardo Rubino, Kishor Trivedi
April 2001

Chapter 1

Introduction

Boudewijn R. Haverkort
Raymond Marie
Gerardo Rubino
Kishor S. Trivedi

1.1 The performability concept

OVER the last two decades there has been an increased interest in the integrated modelling of performance and dependability aspects of computer and communication systems.[1] This so-called *performability modelling and evaluation* has been motivated by the fact that in many modern computer and communication systems, the servicing of jobs can continue, even in the presence of failures. Examples of such systems are fault-tolerant computer systems, parallel computer systems, and distributed computer systems. In all these cases, the failure of a specific component does not necessarily affect the overall service provided by the system, but it does

---

[1]Note that we use the term *dependability* here to denote either reliability or availability. In the original definition of Laprie, security and safety aspects are also included [274]. We do not address these aspects here.

affect the speed, i.e., the performance, at which the system works.

Since more and more applications depend on the correct and timely operation of systems of the above classes (airline reservation systems, banking systems, aircraft control systems, integrated office automation systems, embedded systems of various kinds, etc.), the analysis of their performance in the presence of failure (and repairs) is of major importance. Moreover, many users of these types of systems, such as telephone operating companies and banks, often require that suppliers give evidence that their systems can provide a particular quality of service over a certain time span, even in the presence of failures.

With the integration of computer systems and telecommunications the *quality of service* (QoS) concept as often used in recommendations by standardization bodies such as ITU-T[2] will become more important as an instrument to qualify systems. The QoS measures must reflect the combined influence of dependability and performance associated factors (Recommendation G.106) [66]. It appears that the concept of performability can be used to handle such quality of service-like measures, thus making it even more important for the future [319]. In a similar way, in embedded systems, the timing-related issues are an integrated part of the system specification and formal proofs are required to certify that the system is doing what it should do, in a timely and dependable way.

For the above reasons performability modelling and evaluation has received much attention, in both the research and the industrial communities. The emphasis, however, has mostly been on the mathematical aspects of performability modelling. Although important and a prerequisite for carrying out any practical performability evaluation, the mathematical aspects are not the only important ones. The issue of constructing performability models in a convenient way is also of utmost importance, hence, there is a need for performability evaluation tools.

Over the last two decades a considerable amount of literature has been published on performability modelling and evaluation (see the extensive list provided at the end of this book). However, these publications have appeared in a wide variety of journals and conference proceedings. Furthermore, existing books on either performance or dependability evaluation address the issue of performability only to a limited extent, if at all. This

---

[2]ITU-T (International Telephone and Telegraph Consultative Committee) is the former CCITT.

book tries to bring together the major achievements that have been made in this area over the last 20 years.

In order to ease the reading of this book, we introduce some basic terminology and notation in Section 1.2, especially as far as the most common mathematics is concerned. Then, in Section 1.3 we briefly present the remaining nine chapters of the book.

1.2   Mathematical framework

Gracefully degradable computer systems are assumed to be able to perform at various levels of performance. At system start all components are assumed to be operational and the system will operate at maximum performance. When a component fails, the system will reconfigure itself and restart its activities, albeit with degraded performance. Due to the fact that time intervals between failures are in general relatively large, it is usually assumed that the system will be in steady state most of the time between successive reconfigurations and failures.

Let $\mathcal{S}$ denote the set of all possible configurations in which the system can operate. Now, define a continuous time stochastic process $X = \{X(t), t \geq 0\}$ on $\mathcal{S}$, describing the structure of the system at time t. This process is defined in such a way that, knowing $X(t)$, we must be able to know which components are up and which are down at time t. Process X is often referred to as the *structure state process* since it describes the state of the system structure. The (steady-state) performance of the system when in structure state $i \in \mathcal{S}$ is denoted by $r_i = r(i)$. The real function $r : \mathcal{S} \rightarrow \mathbb{R}$ is called *reward rate function* on the state space $\mathcal{S}$.

Often, the reward function r is "specified" using multiple performance analyses in the following way. Suppose that r_i signifies the throughput of some system when its state is i. To obtain its value, we can do a "classical" performance analysis, assuming that the system is perfect (no failures), where the configuration of up and down components is that of state i. The value r_i somehow summarizes then the performance of the system in structure state i. This is much more simple than coping with all possible performance states in every possible structure state, which can lead to very complex models, and thus, to a very costly evaluation process. This decomposition, i.e., the separate analysis of the performance given a particular structure state and the subsequent reward analysis, is normally valid in practice due to time-scale differences between the task processing processes and the failures ones. Arguments similar to those used by Courtois in his book on (near-) complete decomposability [94] are used to prove this validity.

For $i \in \mathcal{S}$, let π_i denote the steady-state probability of residing in state i, and $p_i(t)$ the (transient) probability of residing in i at time t. The following basic measures can be distinguished:

1. The steady-state performability (SSP) is defined as:

$$\text{SSP} = \sum_{i \in \mathcal{S}} \pi_i r_i.$$

   In words, SSP is the expected asymptotic reward.

2. The transient analog of this measure is the transient or point performability (TP):

$$\text{TP}(t) = \sum_{i \in \mathcal{S}} p_i(t) r_i.$$

   In words, TP(t) is the expected reward (rate) at time t.

3. When the model has absorbing states, an important measure of interest is the mean reward to absorption (MRTA):

$$\text{MRTA} = \int_0^\infty r_{X(s)} ds.$$

   In words, MRTA is the expected cumulated reward up to absorption.

4. Finally, the cumulative performability (CP) is defined as:

$$\text{CP}(t) = \int_0^t r_{X(t)} ds.$$

   Note that CP(t) is in general a random variable. Its distribution

$$F_{\text{CP}}(t, y) = \Pr\{\text{CP}(t) \le y\}$$

   is called the *performability distribution*.

The usual situation is that X is a continuous-time Markov chain. This means that the process $\{r_{X(t)}, t \geq 0\}$ is a Markov reward process. The MRTA can only be defined for models with absorbing states (see below). It should be noted that there is some debate about whether all the above measures are performability measures. Some authors require performability measures to be at least time dependent, thus excluding the steady-state performability measures. Originally, only the distribution of the cumulative reward was addressed and called *the* performability [315, 316, 319]. Since the determination of this distribution is often difficult or costly, often only its moments are considered.

If one chooses $r_i = 1$, if the system is up in state $i \in S$, and $r_i = 0$ otherwise, then $E[\mathrm{CP}(t)]$ is simply the cumulative uptime. The fraction $E[\mathrm{CP}(t)/t]$ then equals the interval availability.

The basic model discussed so far is a so-called *rate-based* reward model which means that when residing in a particular structure state $i \in S$ at time t, the system performs with rate r_i. The rates, however, may also depend on the global time t, thus having a reward rate function $r_i(t)$ for every state $i \in S$. It is also possible to address *impulse-based* reward models. With these models a *reward impulse function* $r : S \times S \rightarrow \mathbb{R}$ is defined which associates a reward $r_{i,j} = r(i, j)$ to the transition from state $i \in S$ to state $j \in S$. Every time a transition from state i to state j takes place, the cumulatively obtained reward increases instantaneously with $r_{i,j}$ units. Of course these rewards may also depend on the global time, thus having transition reward functions $r_{i,j}(t)$. In general, combinations of these four possibilities may coexist. For an overview of various Markov reward models the reader is referred to Howard [236]. Other authors have considered various extensions such as non-homogeneous Markov reward models, semi-Markov reward models and Markov regenerative reward models. Combinatorial performability models have also been developed. In these cases, the approach consists of considering combinatorial models with multiple-valued states, instead of the classical binary ones, underlying the fact that the system can behave at various performance levels. This, again, combines into a unique model, performance and dependability aspects of the same system (see [20]).

To finish this introduction, let us illustrate with a simple example the difference between a performability evaluation and a simpler performance or dependability evaluation. Consider a dual-processor system where both

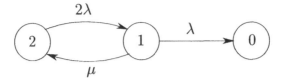

Figure 1.1: State-transition diagram of the dual-processor system

processors work in parallel on different tasks. When one of the processors has a breakdown, the remaining operational unit puts the failed one back in operation after a random time which is exponentially distributed with rate μ. If the remaining unit itself fails before the end of the repair process, the whole system is down and no more repairs are possible without an external action. Assuming exponential operational periods (with rate λ) and the usual independence hypothesis, the process X defined by $X(t) =$ "number of operational processors at time t", valued on $\{2, 1, 0\}$, is Markov homogeneous with an absorbing state (state 0). The corresponding state transition diagram is given in Figure 1.1.

When the system works with only one operational unit, it produces a gain flow of g dollars per unit of time; when it works at maximum capacity of two operational units, its gain flow rate is αg dollars per unit of time, where $1 < \alpha < 2$. Thus, these weights (or reward rates) on the states 2 and 1 (with values αg and g respectively) take into account the effect on the performance of the failures. A measure that ignores this fact is, for instance, the mean total operational time of the system,

$$E\left[\int_0^\infty 1_{(X(t)\neq 0)}dt\right] = \frac{3\lambda + \mu}{2\lambda^2},$$

in which no difference is made between the states in which the system is working with only one or with two processors. An example of an adequate performability measure here is the mean gain obtained during the operational life of the system:

$$E\left[\int_0^\infty \left(\alpha g 1_{(X(t)=2)} + g 1_{(X(t)=1)}\right) dt\right] = g\frac{2\lambda + \alpha(\lambda + \mu)}{2\lambda^2}.$$

The derivation of this measure here is a matter of elementary applied probability; note that the former result is obtained as a special case with $\alpha = 1$ and $g = 1$.

One of the aims of this monograph is to describe the techniques that can be used to deal with the general case and with more complex measures such as the performability distribution.

1.3 Overview of the book

The rest of this book consists of nine chapters, each addressing a specific and important aspect of performability modelling and evaluation.

In Chapter 2 Mitrani addresses a particular class of performability models, namely queueing networks in which the queues are served by unreliable processors. In other words, the processors are subject to failures and repairs while being busy serving jobs. The chapter focuses on the analysis of a single queue, in which case three methods exist to compute detailed performability measures: generating functions, matrix-geometric methods and the spectral expansion method. These three methods are presented and compared with respect to their efficiency. Extensions to the case where the failures and repairs are triggered by job arrivals or departures, and the case where failures and repairs at different processors are dependent on each other, are presented as well.

The important uniformization technique is presented in Chapter 3 by de Souza e Silva and Gail. Uniformization is a transformation allowing the solution in discrete time of a problem whose setting is in terms of a continuous time Markov chain. It has been mainly designed to work with distributions, that is, to compute the probability distribution associated with the process state. The approach has three important benefits: it leads to stable algorithms (it involves only additions and multiplications), it allows control of the numerical errors in the computations, and there is a nice probabilistic interpretation of the transformation. The authors describe the basic technique and present some extensions that have been proposed to it, either to improve the efficiency of the obtained algorithms or to be able to deal with new measures. For instance, techniques proposed to handle non-homogeneous processes, or allowing one to generate the transition probability matrix dynamically, in parallel with the evolution of the computations, are discussed. It also describes how to obtain expected values and not only distributions, and how to combine uniformization with other solution techniques.

For a specific class of Markov-reward models, it is possible to express the performability distribution or its moments explicitly. This issue of so-called closed-form solutions for performability is addressed in Chapter 4 by Donatiello and Grassi. They give an overview of when such solutions exist, as well as discuss their form. They also show that the main equations to derive these closed-form solutions are generalizations of the well-known Markov renewal equation. The closed-form solutions they arrive at take the form of sums of products of exponential terms of which the coefficients can be derived recursively from the parameters in the Markov-reward models. The relation of the closed-form solutions with some of the solutions that are obtained with uniformization in Chapter 3 is explicitly addressed. The chapter is concluded with an overview of all the known methods for closed-form performability, discussing their computational complexity and the constraints on their applicability.

In Chapter 5, Pattipati, Mallubhatla, Gopalakrishna and Viswanatham view performability models as partial differential equations of the hyperbolic type. This view leads to use of the important set of tools available in classical analysis to solve the problems considered in this monograph. The goal of the chapter is to discuss the various techniques that can be used to obtain closed-form solutions to the reward equations formalizing basic performability problems, and to derive efficient numerical schemes to perform the computations. The authors also discuss some important properties of the solutions to these equations, that are not only needed to derive some of the proposed techniques, but also interesting in their own right. Another important topic surveyed in this chapter is the derivation of asymptotic properties of the solutions, again obtained by means of the analysis of the mathematical properties of the hyperbolic differential systems.

Given a complete performability model, the exact computation of the steady-state and transient performability measures is a laborious task. Therefore, in Chapter 6, van Dijk addresses the question of how to avoid this laborious task of computing steady-state and transient probabilities by addressing slight adaptations of the model that can be solved more easily, and that give provable upper and/or lower bounds for the measures of interest in the exact model. In particular, van Dijk proposes methods (i) to compute error bounds on the results obtained with the simplified models, (ii) to make comparisons between the exact and the simplified models in order to decide whether the simplified model yields an upper or lower bound on the measure of interest, and (iii) to give bounds for transient performability measures on the basis of steady-state measures (which might be easier to compute).

In Chapter 7 Bobbio and Telek present a user-oriented view towards performability modelling. They address the completion time of a specific task, of which the service demand is specified probabilistically, on a machine that is subject to failures and repairs. The way in which components in the system may fail and can be repaired is also specified stochastically. In particular, the authors address the way in which the completion time (which is fully characterized by a distribution) of a task changes when the method of handling work on the task after the occurrence of failures is varied: is the work completely lost, or can the work continue from the point prior to the failure? In the last part of their chapter the authors show how the completion time problem can be described nicely using recent advances on non-Markovian stochastic Petri nets.

In Chapter 8 Shahabuddin discusses one of the main issues when the chosen evaluation technique is simulation. This method is a powerful one, since it can handle very general models and it allows the users to evaluate arbitrarily complex measures. Exact numerical techniques may suffer from a combinatorial explosion of the state space, and in many situations they simply cannot provide the values of the metrics of interest. In principle, simulation can deal with virtually any model, but the approach has its own complexity problems. The most important one is the fact that when the event of interest is rare, that is, when the key phenomenon in the system happens rarely (for instance, a catastrophic failure), standard simulation methods may fail because of the excessive computing time necessary to observe such events often enough in order to obtain precise estimations. This means that specific techniques must be used. Shahabuddin describes some of these techniques, namely importance sampling and importance splitting, which in some cases allow one to evaluate measures concerning rare events very efficiently.

In Chapter 9 Meyer and Sanders address the automated derivation of the stochastic process (typically a continuous-time Markov chain) for performability evaluation from a high-level model specification. In particular, they discuss how the actual form (and size) of that stochastic process depends on the performability measure of interest. Their goal is to be able to generate exactly that stochastic process that supports the evaluation of the requested measures, and nothing more (larger) than that. Although they have developed their so-called measure-driven performability model generation in the context of stochastic activity networks, the methods presented in this chapter also apply to other specification formalisms. A large number of detailed examples illustrate their methods.

In Chapter 10 Haverkort and Trivedi address the important issue of having software tools available that help in the construction and the subsequent solution of performability models. In particular, after a brief description of some typical performability measures, the authors describe a general architecture for performability modelling tools. The authors stress the importance of a hierarchical modelling environment in which mathematical details of the solution algorithms are hidden from the tool users. Then, in some detail, a structured overview of 17 tools for performability modelling and evaluation is given. For each tool the authors provide references and implementation details, the supported model class and evaluation techniques, the input and output format and user-interfaces, and the way the tool fits with the proposed general tool architecture.

The book ends with an integrated list of 480 bibliographic items, a glossary and an index.

Chapter 2

Queues with Breakdowns

Isi Mitrani

This chapter surveys several models where one or more queues are served by unreliable processors. The breakdowns and repairs may occur independently, or they may be controlled by some general Markovian environment. A breakdown may simply interrupt a service, or it may result in the loss of an entire queue. The emphasis in all cases is on exact analysis. Steady-state distributions and performance measures are obtained by means of generating functions, the matrix-geometric method or spectral expansion. The relative merits of these techniques are discussed.

2.1 Introduction

THE problem of modelling the behaviour of queueing systems subject to interruptions of service has long been recognized as interesting and important, and has received considerable attention in the literature. Originally, the motivation for studying such models came from the field of manufacturing. This is reflected in the terminology of many early papers, where the servers are referred to as "machines" (e.g., in the *machine interference* model, a finite number of machines, when broken, compete for the attention of a finite number of repairmen; the tasks performed by the operative machines are ignored). On the other hand, most of the more recent research has been motivated by problems in computing and communications; hence the machines have become "processors".

The first infinite-state model with breakdowns that appeared in the literature involved a single server queue where the server goes through alternating periods of being operative and inoperative [17, 153, 459, 478]. A significant advance was achieved later by analyzing a Markovian queueing model with a number of identical parallel servers, each of which breaks down and is repaired independently of the others [323, 351]. That "multiprocessor" queue, together with three different methods that can be applied to its analysis, is described in Section 2.2. The restrictive assumptions of the multiprocessor model can be relaxed in several directions [324]:

- Breakdowns and repairs may depend on the number of operative processors and may occur in several of them simultaneously.

- Breakdowns and/or repairs may be triggered by job arrivals and/or departures.

- Jobs may arrive and/or leave in batches of fixed or variable size.

- The processors may have different service, breakdown and repair characteristics.

These generalizations are discussed in Section 2.3. The next step is to consider systems with more than one queue, i.e., queueing networks. A model where jobs from a common source can be routed through several possible gateways, each with its own queue and server, and with catastrophic breakdowns entailing losses and re-routing, was examined in [327]. This is presented in Section 2.4. Some directions for future study are addressed in Section 2.5.

2.2 The multiprocessor queue with independent interruptions

Consider an M/M/N queue with independent random breakdowns and repairs. Jobs arrive in a Poisson stream at rate σ and join a single, unbounded queue. Their required service times are i.i.d. random variables distributed exponentially with parameter μ. There are N identical parallel processors, each of which goes through alternating periods of being operative and inoperative, independently of the others. The operative periods are i.i.d. random variables distributed exponentially with parameter ξ, and the inoperative ones are i.i.d. random variables distributed exponentially with parameter η. Jobs are taken for service from the front of the queue, one at a time, by available operative processors. A job cannot occupy more than one processor simultaneously, and a processor cannot serve more than one job simultaneously. No operative processor can be idle if there are jobs waiting to be served. If a service is interrupted by a processor breakdown, i.e., an end of an operative period, then the relevant job is returned to the front of the queue. When an operative processor becomes again available for it, the service is resumed from the point of interruption; there is no switching overhead.

The evolution of the system state is represented by the irreducible Markov process $X = \{(I(t), J(t)) \; ; \; t \geq 0\}$, where $I(t)$ is the number of operative processors and $J(t)$ is the number of jobs present at time t. Denote the steady-state distribution of that process by

$$p_{i,j} = \lim_{t \to \infty} P(I(t) = i, J(t) = j) \; ; \; i = 0, 1, \ldots, N \, , \, j = 0, 1, \ldots \, . \quad (2.1)$$

The marginal distribution of the number of operative processors is easily seen to be binomial:

$$p_{i,\cdot} = \sum_{j=0}^{\infty} p_{i,j} = \binom{N}{i} \left(\frac{\eta}{\xi + \eta}\right)^i \left(\frac{\xi}{\xi + \eta}\right)^{N-i} , \; i = 0, 1, \ldots, N \, . \quad (2.2)$$

Hence, the processing capacity of the system, which is defined as the average number of operative processors, is equal to

$$E(I) = \frac{N\eta}{\xi + \eta} \, . \quad (2.3)$$

The process X is ergodic if, and only if, the offered load is less than the processing capacity:

$$\frac{\sigma}{\mu} < \frac{N\eta}{\xi + \eta} \, . \quad (2.4)$$

There are three methods for determining the probabilities $p_{i,j}$, and hence various system performance measures. They involve *generating functions*, *matrix-geometric representation* and *spectral expansion*, respectively. These will be discussed and compared in the following subsections.

2.2.1 Generating functions

Define

$$g_i(z) = \sum_{j=0}^{\infty} p_{i,j} z^j \ , \ i = 0, 1, \ldots, N \ . \tag{2.5}$$

Using a standard technique, the steady-state balance equations for the probabilities $p_{i,j}$ are transformed into a set of $N + 1$ linear equations for the functions $g_i(z)$ (see [323]):

$$A(z)\mathbf{g}(z) = \mathbf{b}(z) \ . \tag{2.6}$$

The coefficient matrix, $A(z)$, is given by:

$$\begin{bmatrix} f_0(z) & -N\eta z & & & \\ -\xi & f_1(z) & -(N-1)\eta z & & \\ & -2\xi z & f_2(z) & -(N-2)\eta z & \\ & \ddots & \ddots & \ddots & \ddots \\ & & & -(N-1)\xi z & f_{N-1}(z) & -\eta z \\ & & & & -N\xi z & f_N(z) \end{bmatrix} , \tag{2.7}$$

where $f_0(z) = \sigma(1 - z) + N\eta$ and $f_i(z) = \sigma z(1 - z) - i\mu(1 - z) + i\xi z + (N - i)\eta z$, $i = 1, 2, \ldots, N$. The vector of free terms has the form $\mathbf{b}(z) = [0, b_1(z), \ldots, b_N(z)]$, where

$$b_i(z) = (z - 1) \sum_{j=0}^{i-1} (i - j)\mu p_{i,j} z^j \ , \ i = 1, 2, \ldots, N \ . \tag{2.8}$$

Then Cramer's rule provides the following expressions for $g_i(z)$:

$$g_i(z) = \frac{D_i(z)}{D(z)} \ , \ i = 0, 1, \ldots, N \ , \tag{2.9}$$

where $D(z)$ is the determinant of $A(z)$ and $D_i(z)$ is the determinant of the matrix obtained from $A(z)$ by replacing its ith column with the vector $\mathbf{b}(z)$.

Note that $\mathbf{b}(z)$ contains $N(N+1)/2$ unknown probabilities. Additional relations for these unknowns are provided by the balance equations for $i = 2, 3, \ldots, N$ and $j = 0, 1, \ldots, i - 1$, and by the normalizing equation. This still leaves a shortfall of $N - 1$ equations. The latter are derived by observing that the right-hand side of (2.9) must be analytic for all z in the interior of the unit disc. That observation is combined with the following fact [323]: the polynomial $D(z)$, which is of degree $2N + 1$, has $2N + 1$ real and positive roots, $z_1, z_2, \ldots, z_{2N+1}$. Moreover, when (2.4) holds, those roots are distinct and, if numbered in ascending order, satisfy

$$0 < z_1 < z_2 < \ldots < z_{N-1} < 1 = z_N < z_{N+1} < \ldots < z_{2N+1} \, .$$

The requirement that the right-hand side of (2.9) should be finite at points $z_1, z_2, \ldots, z_{N-1}$ yields the necessary $N - 1$ equations and enables all unknown probabilities to be determined.

2.2.2   Matrix-geometric solution

The second solution method relies on a representation of the steady-state probabilities in terms of the powers of a matrix [350, 351]. This requires the introduction of the row vectors

$$\mathbf{v}_j = (p_{0,j}, p_{1,j}, \ldots, p_{N,j}) \, , \; j = 0, 1, \ldots \, , \tag{2.10}$$

whose elements are the probabilities of all states with j jobs present in the system. When $j \geq N$, these vectors can be shown to satisfy a difference equation of order 2 with constant coefficients:

$$\mathbf{v}_j Q_0 + \mathbf{v}_{j+1} Q_1 + \mathbf{v}_{j+2} Q_2 = \mathbf{0} \, , \; j = N, N + 1, \ldots \, . \tag{2.11}$$

Here, Q_0 is a diagonal matrix containing the arrival rate: $Q_0 = \sigma I$, where I is the identity matrix of order $N+1$; Q_2 is the diagonal matrix containing the possible service rates: $Q_2 = diag[0, \mu, 2\mu, \ldots, N\mu]$. The matrix Q_1 is defined as follows. Denote

$$B = \begin{bmatrix} 0 & N\eta & & & \\ \xi & 0 & (N-1)\eta & & \\ & 2\xi & 0 & \ddots & \\ & & \ddots & \ddots & \eta \\ & & & N\xi & 0 \end{bmatrix} \, .$$

It is also convenient to denote by $\Delta(A)$ the diagonal matrix containing the row sums of any given matrix A. Then $Q_1 = B - \Delta(B) - \Delta(Q_0) - \Delta(Q_2)$.

Note that the diagonal elements of $\Delta(B)$, $\Delta(Q_0)$ and $\Delta(Q_2)$ are the total rates at which the Markov process leaves state (i, j), due to changes in the number of operative servers, job arrivals and job departures, respectively. Equations (2.11) are simply the steady-state balance equations written by means of vectors and matrices. Their solution can be expressed in the form

$$\mathbf{v}_j = \mathbf{v}_N R^{j-N} , \quad j = N, N+1, \ldots , \tag{2.12}$$

where R is the minimal solution of the quadratic matrix equation corresponding to (2.11):

$$Q_0 + RQ_1 + R^2 Q_2 = 0 . \tag{2.13}$$

The vectors $\mathbf{v}_N$, $\mathbf{v}_{N-1}$, $\ldots$, $\mathbf{v}_0$ are determined from the balance equations for $j \leq N$, together with the normalizing equation.

The matrix equation (2.13) is usually solved by an iterative procedure recommended by Neuts:

$$R(n+1) = [-Q_0 - R(n)^2 Q_2]Q_1^{-1} . \tag{2.14}$$

The starting point can be $R(0) = 0$. The number of iterations depends on the model parameters and on the desired accuracy.

2.2.3 Spectral expansion

The third solution method also relies on solving the difference equation (2.11), but uses a different approach [325, 324, 350]. Associated with (2.11) is a matrix polynomial

$$Q(\lambda) = Q_0 + Q_1\lambda + Q_2\lambda^2 . \tag{2.15}$$

This is known as the *characteristic polynomial* of (2.11). Denote its eigenvalues in the interior of the unit disc, and the corresponding left eigenvectors, by λ_k and ψ_k respectively. Suppose that there are c such eigenvalues and eigenvectors. In other words, these are quantities which satisfy

$$det[Q(\lambda_k)] = 0 , \quad |\lambda_k| < 1 , \quad \psi_k Q(\lambda_k) = \mathbf{0} , \quad k = 1, 2, \ldots, c . \tag{2.16}$$

Then the solution of (2.11) can be expressed in the form

$$\mathbf{v}_j = \sum_{k=1}^{c} x_k \psi_k \lambda_k^j , \quad j = N, N+1, \ldots , \tag{2.17}$$

where x_k $(k = 1, 2, \ldots, c)$ are arbitrary constants.

The coefficients in the spectral expansion (2.17), and the vectors $\mathbf{v}_{N-1}$, $\ldots$, $\mathbf{v}_0$ are again determined from the balance equations for $j \leq N$, together with the normalizing equation. However, in order that the number of available equations is equal to the number of unknowns, it is necessary that the number, c, of eigenvalues in the interior of the unit disc is equal to $N + 1$. It can in fact be shown that when the process is ergodic, there are exactly $N + 1$ eigenvalues inside the unit disc. Moreover, they are all real, positive and simple.

The computation of λ_k and ψ_k is perhaps best carried out by reducing the quadratic eigenvalue–eigenvector problem

$$\psi[Q_0 + Q_1\lambda + Q_2\lambda^2] = \mathbf{0} \tag{2.18}$$

to a linear one of the form $\mathbf{y}Q = \lambda\mathbf{y}$, where Q is a matrix whose dimensions are twice as large as those of Q_0, Q_1 and Q_2. The latter problem is normally solved by applying various transformation techniques. Efficient routines for that purpose are available in most numerical packages.

To carry out the linearization, divide (2.18) by σ (remember that $Q_0 = \sigma I$), perform a change of variable $\lambda \to 1/\lambda$, and multiply by λ^2. Then (2.18) becomes

$$\psi[R_2 + R_1\lambda + I\lambda^2] = \mathbf{0}, \tag{2.19}$$

where $R_1 = Q_1/\sigma$ and $R_2 = Q_2/\sigma$ (these should not be confused with Neuts' matrix R). The relevant eigenvalues are now the ones *outside* the unit disc. Next, by introducing the vector $\varphi = \lambda\psi$, equation (2.19) can be rewritten in the equivalent linear form

$$[\psi, \varphi] \begin{bmatrix} 0 & -R_2 \\ I & -R_1 \end{bmatrix} = \lambda[\psi, \varphi]. \tag{2.20}$$

2.2.4 Comparison

A few words are in order concerning the numerical efficiency of the three methods. Both the generating function and the spectral expansion methods rely on the computation of a set of eigenvalues. However, the spectral expansion method yields the steady-state probabilities directly, without having to invert a generating function. This may be an important advantage, e.g., if one is interested in determining percentiles of the queue size distribution.

The trade-offs between the matrix-geometric and the spectral expansion methods are those between the iterative solution of the quadratic matrix equation (2.13), and the calculation of the eigenvalues and eigenvectors of $Q(\lambda)$. There is strong evidence [324] that the latter is considerably faster than the former when the system is heavily loaded. This is due to the fact that the number of iterations required to compute the matrix R increases without bound when the system approaches saturation. On the other hand, the computation time for the eigenvalues and eigenvectors is not affected by the load.

The overall computational complexity of determining all eigenvalues and eigenvectors of an $N \times N$ matrix is $O(N^3)$. Moreover, that complexity is more or less independent of the parameters of the model. The accuracy of the spectral expansion solution is limited only by the precision of the numerical operations. Using double precision arithmetic, the performance measures computed by this method are correct to about 10 significant digits.

If the matrix R is computed approximately by an iterative procedure such as (2.14), then the criterion for termination would normally include some trade-off between computation time and accuracy. The situation is illustrated in Table 2.1, where a 10-server model is solved by both methods. The procedure (2.14) is terminated when

$$\max_{i,j} |R_{n+1}(i,j) - R_n(i,j)| < \epsilon \, ,$$

for a given value of ϵ. The performance measure that is computed is the average queue size, $E(J)$. Of course the value obtained by spectral expansion (column 4) does not depend on ϵ. The offered load is varied by increasing the job arrival rate, σ (the system saturates when $\sigma = 6.6666...$).

The table confirms that when ϵ decreases, the matrix-geometric solution approaches the solution by spectral expansion. However, it is important to observe that the accuracy of R is not related in any obvious way to the accuracy of $E(J)$. Thus, taking $\epsilon = 10^{-6}$ yields an answer whose relative error is 0.0004% when $\sigma = 3$, 0.06% when $\sigma = 6$, and 6.3% when $\sigma = 6.6$. Another important aspect of the table is that, for given ϵ, the number of iterations required to compute R increases with σ.

Thus, a comparison between the run times of the spectral expansion and the matrix-geometric algorithms shows that the former is independent of the parameters (and the question of accuracy does not arise), while the

Table 2.1: Trade-off between accuracy and complexity

ϵ	iter.	$E(J)$ (m. geom.)	$E(J)$ (sp. exp.)	% difference
		$\sigma = 3.0$		
10^{-3}	29	5.1750728791	5.1997104203	0.4738252555
10^{-6}	93	5.1996865270	5.1997104203	0.0004595110
10^{-9}	158	5.1997103968	5.1997104203	0.0000004521
10^{-12}	223	5.1997104203	5.1997104203	0.0000000004
		$\sigma = 6.0$		
ϵ	iter.	$E(J)$ (m. geom.)	$E(J)$ (sp. exp.)	% difference
10^{-3}	77	35.038255532	50.405820557	30.487679509
10^{-6}	670	50.372679585	50.405820557	0.065748305
10^{-9}	1334	50.405787235	50.405820557	0.000066108
10^{-12}	1999	50.405820524	50.405820557	0.000000066
		$\sigma = 6.6$		
ϵ	iter.	$E(J)$ (m. geom.)	$E(J)$ (sp. exp.)	% difference
10^{-3}	77	58.19477818	540.46702456	89.23250160
10^{-6}	2636	506.34712584	540.46702456	6.31303986
10^{-9}	9174	540.42821366	540.46702456	0.00718099
10^{-12}	15836	540.46698572	540.46702456	0.00000719

latter is influenced by both the offered load and the desired accuracy. For a heavily loaded system, the matrix-geometric run time may be many orders of magnitude larger than that of the spectral expansion. The extra time is almost entirely taken up by the iterative procedure for calculating the matrix R. It seems clear that, even when the accuracy requirements are modest, the computational demands of that procedure have a vertical asymptote at the saturation point.

2.2.5 Asymptotics

The asymptotic behaviour of the system under heavy load, i.e., when the parameters are such that the average queue size tends to infinity, is also of interest. Under those conditions, the limiting state distribution has a considerably simpler form than in the general case, and is easier to compute. That asymptotic distribution can then be used as an approximate solution for a heavily loaded system.

For a given number of processors, there are two ways of increasing the load. One is to let the average amount of work entering the system per unit time (the ratio of arrival rate to service rate) approach the available service capacity (the average number of operative processors), while keeping the breakdown and repair rates bounded away from zero. This is usually referred to as the *heavy traffic* limit. It can be demonstrated (see [326]) that in this case the limiting distribution of the number of jobs in the system, appropriately normalized, is exponential. Moreover, the number of operative processors and the number of jobs in the system are asymptotically independent of each other.

The precise form of that result is as follows. Let $\rho = \sigma/[\mu E(I)]$, with $E(I)$ given by (2.3), be the system load; $\rho < 1$ is the ergodicity condition. Denote also by λ_{N+1} the largest eigenvalue in (2.16). That eigenvalue approaches 1 when $\rho \to 1$. Then the limiting joint distribution of the number of operative processors and the number of jobs in the system normalized by multiplying it with $1 - \lambda_{N+1}$ has the form

$$\lim_{\rho \to 1} P[I = i , \ (1 - \lambda_{N+1})J \leq x] = p_{i,\cdot}(1 - e^{-x}) , \ x \geq 0 , \ i = 0, 1, \ldots, N,$$

where $p_{i,\cdot}$ is given by (2.2). Other normalizations of J are also possible and give similar results, perhaps with different means of the exponential distribution.

The other limiting process is to let the breakdown and repair rates approach zero, while keeping their ratio (as well as the other parameters) fixed. Such a limit maintains the ratio of offered load to service capacity and so apparently it does not conform to the notion of heavy load. However, when the average lengths of the operative and inoperative periods are increased in a fixed proportion, the average queue size grows without bound. This reflects a situation likely to occur in practice, where processors break down rarely but take a long time to repair. Now the limiting distribution of a normalized number of jobs in the system can be shown to have a rational Laplace transform with simple poles. The number of jobs and the number of operative processors maintain their dependence in the limit. The details of that development can be found in [326].

2.3   Generalizations

Many of the restrictions of the above model can be relaxed by assuming that the availability of the processors is governed by a general Markovian environment which may also depend on the arrival and departure of jobs. Under these assumptions, multiple simultaneous breakdowns or repairs can be allowed, including ones triggered by job arrivals or departures. Arrival rates may depend on the number of operative processors. Moreover, the arrivals and departures may occur in batches of fixed or random size (in the latter case those sizes should be bounded).

Clearly, there are applications that give rise to models incorporating one or more of the above general features. For example, certain hardware faults, e.g., in buses or power supplies, cause simultaneous service interruptions in a cluster of processors. On the other hand, removing such faults, or replacing groups of inoperative processors, can be treated as simultaneous repairs.

Denote again by $I(t)$ and $J(t)$ the random variables representing the number of operative processors and the number of jobs in the system at time t, respectively. Consider first the case when arrivals and departures occur singly. Then the evolution of the Markov process proceeds according to the following set of possible transitions:

(a) from state (i, j) to state $(k, j + 1)$ $(0 \leq i, k \leq N)$;

(b) from state (i, j) to state (k, j) $(0 \leq i, k \leq N \; ; \; i \neq k)$;

(c) from state (i, j) to state $(k, j - 1)$ $(0 \leq i, k \leq N)$.

There may be arbitrary state-dependency in the instantaneous transition rates up to some threshold, M, of the number of jobs in the system. Beyond that threshold, the instantaneous transition rates do not depend on j (although they may still depend on i). In other words, if we denote the transition rate matrices associated with (a), (b) and (c) by A_j, B_j and C_j respectively (the main diagonal of B_j is zero by definition; also, $C_0 = 0$ by definition), then we have

$$A_j = A \; , \quad B_j = B \; , \quad C_j = C \; , \quad j \geq M \qquad (2.21)$$

(this matrix A and the one introduced in Section 2.2 are not related).

A transition of type (a) represents a job arrival, coinciding perhaps with a change in the number of operative processors. If arrivals are not accompanied by such changes, then the matrices A and A_j are diagonal. For example, if jobs arrive according to a state-dependent Poisson process with instantaneous rate σ_i when there are i operative processors, then

$$A = A_j = diag[\sigma_0, \sigma_1, \ldots, \sigma_N] \,, \ j = 0, 1, \ldots . \tag{2.22}$$

Transitions (b) correspond to changes in the number of operative processors. Clearly, a large class of different breakdown and repair patterns can be modelled by choosing the matrices B and B_j appropriately. For example, suppose that, in addition to the separate breakdowns, it is possible for all currently operative processors to break down simultaneously. These "global" breakdowns occur at rate ξ_0. Similarly, in addition to the separate repairs, it may be possible for all currently inoperative processors to be repaired simultaneously. These global repairs occur at rate η_N. Then the matrices B_j would have the form

$$B_j = B = \begin{bmatrix} 0 & N\eta & & & \eta_N \\ \xi_0 + \xi & 0 & (N-1)\eta & & \eta_N \\ \xi_0 & 2\xi & 0 & \ddots & \vdots \\ \vdots & & & \ddots & \eta + \eta_N \\ \xi_0 & & & N\xi & 0 \end{bmatrix} . \tag{2.23}$$

A transition of type (c) represents a job departure coinciding with a change in the number of operative processors. If such coincidences do not occur, then the matrices C and C_j are diagonal. In particular, if each operative processor services jobs at rate μ, then

$$
\begin{aligned}
C_j &= diag[0, min(j,1)\mu, min(j,2)\mu, \ldots, min(j,N)\mu] \,, \ j < N \,, \\
C &= diag[0, \mu, 2\mu, \ldots, N\mu] \,, \ j \geq N \,. \tag{2.24}
\end{aligned}
$$

The requirement that all transition rates, and in particular those affecting the reliability of the system, cease to depend on the size of the job queue beyond a certain threshold, is not too restrictive. It does enable the consideration of models where the breakdown and repair rates depend on the number of busy processors. On the other hand, it is difficult to think of applications where those rates would depend on the number of jobs waiting to begin execution. Note that we impose no limit on the magnitude of the threshold M, although it must be pointed out that the larger M is, the greater the complexity of the solution. Of course, if the only dependencies of the various transition rates on j are the ones manifested in equations (2.24) and/or the ones mentioned in this paragraph, then the threshold is equal to the number of processors: $M = N$.

All three solution methods described in Section 2 can be applied to this more general model. In particular, the row vectors $\mathbf{v}_j$ satisfy, beyond the threshold M, a difference equation similar to (2.11):

$$\mathbf{v}_j Q_0 + \mathbf{v}_{j+1} Q_1 + \mathbf{v}_{j+2} Q_2 = \mathbf{0} , \; j = M, M+1, \ldots , \qquad (2.25)$$

where $Q_0 = A$, $Q_1 = B - \Delta(B) - \Delta(A) - \Delta(C)$, and $Q_2 = C$.

The matrix-geometric and the spectral expansion solutions proceed as described in Section 2.2, using the balance equations below the threshold to determine the unknown constants. Note that now there can be non-real eigenvalues in the unit disc. However, if that is the case, then they appear in complex-conjugate pairs. The corresponding eigenvectors are also complex-conjugate. The same must be true for the appropriate pairs of coefficients in the expansion, x_k, in order that the right-hand side of (2.17) be real. To ensure that it is also positive, it seems that the real parts of λ_k, ψ_k and x_k should be positive. Indeed, that is invariably observed to be the case.

In principle, there may also be multiple eigenvalues, which would cause difficulties for the spectral expansion method. However, the likelihood that a real-life problem of this type will exhibit multiple eigenvalues is negligible (if the coefficients of a random polynomial are sampled from a continuous distribution, the probability that it will have multiple roots is 0).

It should be pointed out again that, for heavily loaded systems, the spectral expansion solution is considerably faster than the matrix-geometric one.

The next generalization is to allow batch arrivals and/or departures. This leads to a Markov process where the variable J may jump by arbitrary but bounded amounts in either direction. In other words, we wish to allow transitions

(a) from state (i, j) to state $(k, j + s)$ $(0 \leq i, k \leq N ; 1 \leq s \leq r_1 ; r_1 \geq 1)$;

(b) from state (i, j) to state (k, j) $(0 \le i, k \le N ; i \ne k)$;

(c) from state (i, j) to state $(k, j - s)$ $(0 \le i, k \le N ; 1 \le s \le r_2 ; r_2 \ge 1)$.

Denote by $A_j(s)$, B_j and $C_j(s)$ the transition rate matrices associated with (a), (b) and (c), respectively. There is a threshold M, such that

$$A_j(s) = A(s) \; ; \; B_j = B \; ; \; C_j(s) = C(s) \, , \, j \ge M \; . \tag{2.26}$$

The balance equations are once more reduced to a vector difference equation with constant coefficients, but this time the order of the equation is $r = r_1 + r_2$:

$$\sum_{\ell=0}^{r} \mathbf{v}_{j+\ell} Q_\ell = 0 \, , \, j \ge M \; . \tag{2.27}$$

Here, $Q_\ell = A(r_1 - \ell)$ for $\ell = 0, 1, \ldots r_1 - 1$,

$$Q_{r_1} = B - \Delta(B) - \sum_{s=1}^{r_1} \Delta(A(s)) - \sum_{s=1}^{r_2} \Delta(C(s)) \, ,$$

and $Q_\ell = C(\ell - r_1)$ for $\ell = r_1 + 1, r_1 + 2, \ldots, r_1 + r_2$.

The matrix-geometric solution can be applied by first determining the minimal solution of a matrix equation of order r:

$$\sum_{\ell=0}^{r} R^\ell Q_\ell = 0 \; . \tag{2.28}$$

This can be done by an iterative scheme similar to (2.14):

$$R(n + 1) = [-Q_0 - \sum_{\ell=2}^{r} R(n)^\ell Q_\ell] Q_1^{-1} \, , \tag{2.29}$$

starting with $R(0) = 0$. Then the solution of (2.27) is expressed in the form

$$\mathbf{v}_j = \mathbf{v}_M R^{j-M} \, , \, j = M, M + 1, \ldots \, , \tag{2.30}$$

and the unknown probabilities are obtained as before.

To apply the spectral expansion solution, one forms the characteristic matrix polynomial

$$Q(\lambda) = \sum_{\ell=0}^{r} Q_\ell \lambda^\ell \; . \tag{2.31}$$

The solution above the threshold is expressed in terms of the eigenvalues of (2.31) in the interior of the unit disc, and the corresponding left eigenvectors, as was done in (2.17). The unknown coefficients of the expansion and the probabilities below the threshold are determined from the remaining balance and normalizing equations.

For computational purposes, one can transform the polynomial eigenvalue–eigenvector problem of degree r into a linear one in a similar way as described in Section 2.2. For example, suppose that Q_0 is non-singular. Multiply (2.31) on the right by Q_0^{-1}, perform a change of variable $\lambda \to 1/\lambda$ and then multiply by λ^r. This leads to the eigenvalue–eigenvector problem

$$\psi\,[\sum_{\ell=0}^{r-1} R_\ell \lambda^\ell + I\lambda^r] = 0 , \qquad (2.32)$$

where $R_\ell = Q_{r-\ell}Q_0^{-1}$. Introducing the vectors $\varphi_\ell = \lambda^\ell\psi$, $\ell = 1, 2, \ldots, r-1$, we obtain the equivalent linear form

$$[\psi, \varphi_1, \ldots, \varphi_{r-1}] \begin{bmatrix} 0 & & & -R_0 \\ I & 0 & & -R_1 \\ & \ddots & \ddots & \\ & & I & -R_{r-1} \end{bmatrix} = \lambda[\psi, \varphi_1, \ldots, \varphi_{r-1}] .$$

Another direction in which the model can be generalized is by allowing more complex operative states for the set of processors. For instance, the processors need not be identical, but may have different breakdown, repair and service characteristics. Then an operative state would be defined by specifying the subset of processors that are operative; the variable $I(t)$ would take 2^N possible values, instead of $N+1$. Also, non-exponential (e.g., Coxian) distributions of various random variables may be handled by the method of stages, again at the price of increasing the number of operative states. Note that as the number of operative states increases, so does the size of Neuts' matrix R, and the number of eigenvalues, eigenvectors and unknown constants that has to be determined.

2.4 Several queues coupled by breakdowns

The models considered so far have led to Markov processes whose state spaces were essentially one-dimensional, because there has been just one unbounded queue. Let us now consider a model involving a network of unbounded queues subject to breakdowns and repairs (for more details, see [327]). Even though the topology of this network is rather simple, the problem is by no means trivial.

Jobs arrive into the system in a Poisson stream with rate σ. There are N different servers, each with an associated unbounded queue, to which incoming jobs may be directed. Server i goes through alternating independent operative and inoperative periods, distributed exponentially with means $1/\xi_i$ and $1/\eta_i$, respectively. While it is operative, the jobs in its queue receive exponentially distributed services with mean $1/\mu_i$, and depart upon completion. When a server becomes inoperative (breaks down), the corresponding queue is cleared and all jobs in it, whether waiting or in service, are lost.

Server i is assigned a positive "routing weight" w_i. If, when a new job arrives into the system, the set of operative servers, α, is non-empty ($\alpha \subset \{1, 2, \ldots, N\}$), then the job is placed into one of the queues in α, choosing queue i with probability $q_i(\alpha)$, given by

$$q_i(\alpha) = \frac{w_i}{\sum_{j \in \alpha} w_j} \; , \quad i \in \alpha \; .$$

These routing decisions are independent of each other and of the numbers of jobs in queues. If all servers are broken down at an arrival instant, then the incoming job is lost.

The motivation for this model comes from the field of networking: the jobs are messages generated by some source, and the servers are alternative gateways through which those messages may be routed. Gateways are subject to failures that are sufficiently catastrophic to lose all messages currently accumulated. The source finds out about such failures and, if possible, redirects traffic. The source can be considered to cease generating messages during the periods when all processors are broken.

The system state at time t is described by the pair $X(t) = (\mathbf{I}(t), \mathbf{J}(t))$, where $\mathbf{I}(t) \subset \{1, 2, \ldots, N\}$ is the set of currently operative servers (represented, if not empty, as a vector of indices), and $\mathbf{J}(t)$ is the vector of corresponding queue sizes. More precisely, if $\mathbf{I}(t) = (i_1, i_2, \ldots, i_k)$, then $\mathbf{J}(t)$ has k elements, the first of which is the number of jobs in queue i_1, the second is the number of jobs in queue i_2, etc. Remember that the queues of inoperative servers are necessarily empty, and so their sizes need not be specified. If $\mathbf{I}(t)$ is the empty set, $\emptyset$, then $\mathbf{J}(t)$ is absent.

The above assumptions ensure that $\{X(t)\ ,\ t \geq 0\}$ is an irreducible Markov process. Moreover, it is intuitively obvious that, no matter how heavy the traffic, the process is ergodic. This is due to the fact that queues are emptied from time to time as result of server breakdowns, and so their sizes cannot grow without bound.

The object of greatest theoretical importance in connection with this model is the joint stationary distribution of the set of operative servers and the numbers of jobs in the corresponding queues. To find that distribution one has to solve a non-separable multidimensional Markov process, which is an intractable problem in general. However, it is possible to determine the joint distribution in the case $N = 2$, by a reduction to a Dirichlet boundary value problem on a circle [327].

On the other hand, the performance measures of practical interest are mainly concerned with averages, e.g. the average number of jobs present at a given server or the total average number of jobs lost per unit time. To calculate such performance measures, it is enough to determine the marginal queue size distributions. This problem can be solved, at least in principle, for arbitrary N. In particular, the ability to compute mean queue sizes implies that one can tackle various optimization problems. For instance, how should the input stream of jobs be split among the possible routes in order to minimize the total average number of losses per unit time?

The marginal distribution of the operative servers is quite straightforward: for every subset, $\alpha \subset \{1, 2, \ldots, N\}$, the probability, p_α, that the servers in α are operative and the ones in $\bar{\alpha}$ are broken (where $\bar{\alpha}$ is the complement of α with respect to $\{1, 2, \ldots, N\}$) is given by

$$p_\alpha = \prod_{i \in \alpha} \frac{\eta_i}{\xi_i + \eta_i} \prod_{i \in \bar{\alpha}} \frac{\xi_i}{\xi_i + \eta_i} \ , \quad \alpha \subset \{1, 2, \ldots, N\} \ , \tag{2.33}$$

where an empty product is by definition equal to 1. These expressions follow from the fact that servers break down and are repaired independently of each other.

Consider, without loss of generality, the marginal distribution of queue 1. Given a subset, $\alpha \subset \{1, 2, \ldots, N\}$, of which 1 is a member, let $p_\alpha(n, \cdot)$ be the equilibrium probability that the servers in α are operative, the ones in $\bar{\alpha}$ are broken down, and there are n jobs in queue 1. For every value of n, there are 2^{N-1} such probabilities (corresponding to the subsets that include 1), and their sum is equal to the equilibrium probability that there are n jobs in queue 1 and server 1 is operative.

By summing the balance equations of the Markov process over all state vectors in which the size of queue 1 is n, and performing the appropriate cancellations, one obtains a set of linear equations for $p_\alpha(n, \cdot)$. These can then be transformed into equations for the generating functions, $g_\alpha(z)$, of the marginal distribution of the number of jobs in queue 1 and the set of operative processors, α.

Knowledge of the functions $g_\alpha(z)$, together with the fact that queue 1 is empty when server 1 is inoperative, yields the generating function of the marginal distribution of queue 1, $h_1(z)$:

$$h_1(z) = \frac{\xi_1}{\xi_1 + \eta_1} + \sum_{1 \in \alpha \subset \{1,2,\dots,N\}} g_\alpha(z) . \tag{2.34}$$

A similar development would lead to the generating function of the marginal distribution of queue i, $h_i(z)$, for $i = 2, 3, \dots, N$. Various performance measures can then be obtained. For example, the total average number of jobs that are lost per unit time, L, is given by

$$L = \sigma p_{\{\emptyset\}} + \sum_{i=1}^{N} \xi_i h_i'(1) . \tag{2.35}$$

Another quantity of interest is the average number, m_i, of successful jobs in queue i. A successful job is one that manages to complete its service, rather than being lost as a result of a breakdown. Note that, if there are n jobs present in queue i, then the probability that j of them are successful is equal to $[\mu_i/(\mu_i + \xi_i)]^j[\xi_i/(\mu_i + \xi_i)]$ for $j = 0, 1, \dots, n-1$, and to $[\mu_i/(\mu_i + \xi_i)]^n$ for $j = n$. From this we can derive a simple relation between the generating function of the number of successful jobs in queue i, $s_i(z)$, and the marginal generating function $h_i(z)$:

$$s_i(z) = \frac{\mu_i(1 - z)}{\xi_i + \mu_i(1 - z)} h_i\left(\frac{\mu_i z}{\mu_i + \xi_i}\right) + \frac{\xi_i}{\xi_i + \mu_i(1 - z)} .$$

Hence, the average number of successful jobs in queue i is given by

$$m_i = s_i'(1) = \frac{\mu_i}{\xi_i}\left[1 - h_i\left(\frac{\mu_i}{\mu_i + \xi_i}\right)\right] . \tag{2.36}$$

The overall average response time, W, of a successful job is obtained by an appeal to Little's theorem:

$$W = \frac{1}{\sigma - L} \sum_{i=1}^{N} m_i \ . \tag{2.37}$$

So, the problem is to determine $g_\alpha(z)$. The approach is somewhat similar to the one described in Section 2.2. The set of equations for $g_\sigma(z)$ is written in matrix form:

$$A(z)\mathbf{g}(z) = \mathbf{b}(z) \ , \tag{2.38}$$

where the elements of $\mathbf{g}(z)$ are the generating functions $g_\alpha(z)$, those of $\mathbf{b}(z)$ are the free terms, and $A(z)$ is the $2^{n-1} \times 2^{n-1}$ coefficient matrix. Then the solution is given by Cramer's rule:

$$g_\alpha(z) = \frac{D_\alpha(z)}{D(z)} \ , \quad 1 \in \alpha \subset \{1, 2, \ldots, N\} \ , \tag{2.39}$$

where $D(z)$ is the determinant of $A(z)$ and $D_\alpha(z)$ is the determinant of the matrix obtained from $A(z)$ by replacing its α-column with the vector $\mathbf{b}(z)$. Unfortunately, the latter vector contains 2^{N-1} unknown probabilities, $g_\alpha(0)$. Additional equations for those probabilities are provided by using the roots of $D(z)$ inside the unit disc. $D_\alpha(z)$ must vanish at those roots. A requirement of the form $D_\alpha(z) = 0$ provides a linear equation for the unknown probabilities. Moreover, if z is a root of multiplicity k, then $k - 1$ derivatives of $D_\sigma(z)$ must vanish, yielding $k - 1$ further equations.

In order to determine 2^{N-1} unknowns, $D(z)$ must have the same number of roots inside the unit disc. This proposition can be proved formally in the case $N = 2$, but is supported only intuitively and numerically in the general case.

In [327] there is some numerical experimentation aimed at examining the optimal splitting of the input stream among two gateways. The results are not always intuitive. For example, under certain conditions, it is best to direct a larger fraction of the incoming jobs towards the server which breaks down more frequently. The problem of optimal routing clearly needs further study.

2.5   Conclusion

The models that have been described here are among the few queueing models involving breakdowns and repairs that can be solved exactly. Yet there are many complex systems whose combined performance and reliability characteristics are of interest. Some obvious examples are queueing networks with general topology and routing matrix, and unreliable nodes. After a breakdown, there may or may not be job losses and re-routing. It is quite likely that such general models will not lend themselves to exact analysis. It would then be desirable to develop approximate solutions and to study their accuracy. This appears to be a fruitful area of further research.

Another interesting problem is the asymptotic behaviour of large networks, or multiprocessors, with "almost reliable" servers. That is, the number of servers grows without bound and their breakdown rate tends to zero, such that the product of the two remains bounded. Such a model would reflect the situation in many current systems, where individual components break down rarely but their number is large.

Chapter 3

The Uniformization Method in Performability Analysis

Edmundo de Souza e Silva
H. Richard Gail

During the last few years a technique to find transient solutions for Markov models, called uniformization or randomization, has become widely used. Reasons such as the simplicity of the basic result, numerical robustness and the probabilistic interpretation which leads to many extensions, help to explain the popularity of the method. We survey the foundations of the technique and describe several extensions proposed in the literature. We present algorithms that are useful in calculating various performability measures.

3.1 Introduction

UNIFORMIZATION, randomization or, as it has been recently called, Jensen's method, is a widely used technique in performance and dependability analysis and, in particular, performability analysis (see [319, 320] for surveys on performability). The technique was proposed by Jensen in 1953 [242], and it has become very popular in the last few years for obtaining transient measures of Markov models. Several books and articles now cover the basic ideas of the method, e.g. [87, 106, 193, 194, 198, 390] (see also [110] for a recent survey on transient analysis), and a large number of papers have been published which use the foundations of the technique to solve a variety of problems, from performance and availability to performability related problems, e.g. [103, 104, 105, 180, 190, 191, 219, 310, 311]. It has also been used together with simulation, e.g. [353, 431].

The method involves transforming a continuous-time Markov chain into a discrete-time analog, and transient solutions are obtained by working with a problem in discrete time that is particularly amenable to numerical calculation. Consider a continuous-time Markov chain with irreducible infinitesimal generator $\mathbf{Q}$. Directly solving the Kolmogorov backward equations of a continuous-time chain for $\mathbf{p}(t)$, the state probability vector at time t, gives

$$\mathbf{p}(t) = \mathbf{p}(0)e^{\mathbf{Q}t} = \mathbf{p}(0) \sum_{n=0}^{\infty} \frac{t^n}{n!} \mathbf{Q}^n, \qquad (3.1)$$

where $\mathbf{p}(0)$ is the initial state distribution. Therefore, (3.1) could be used to evaluate $\mathbf{p}(t)$ numerically. However, computations with (3.1) suffer from severe roundoff errors due to the negative diagonal elements in $\mathbf{Q}$.

Let Λ be at least as large as the maximum absolute value of the diagonal entries of $\mathbf{Q}$, and define the matrix $\mathbf{P} = \mathbf{I} + \mathbf{Q}/\Lambda$, where $\mathbf{I}$ is the identity matrix. Then, we have

$$\mathbf{p}(t) = \mathbf{p}(0)e^{(\mathbf{P}-\mathbf{I})\Lambda t} = \mathbf{p}(0) \sum_{n=0}^{\infty} e^{-\Lambda t} \frac{(\Lambda t)^n}{n!} \mathbf{P}^n. \qquad (3.2)$$

Since Λ is chosen so that the negative diagonal elements of $\mathbf{Q}/\Lambda$ have absolute value at most 1, then $\mathbf{P}$ is a stochastic matrix.

Equation (3.2) is the basic equation of the uniformization method, and it is surprisingly simple. In fact, there are many reasons that explain the popularity of the method, besides its simplicity. These include the beneficial numerical properties of (3.2) and the possibility of easily bounding the solution when the infinite series in (3.2) is truncated. Also, if $\mathbf{P}$ is sparse, which is true for matrices of many models, considerable savings are obtained in the computation of (3.2). But perhaps the biggest advantage of the technique is its probabilistic interpretation, which is discussed in the next section.

In what follows we survey the basics of the uniformization technique together with a few extensions, and address several issues related to performance/availability analysis. Although we discuss uniformization methods applicable to transient analysis in general, we do not cover all of the extensions of the technique. Instead we concentrate on the topics more closely related to performability.

In Section 3.2 we formally introduce the uniformization method and highlight various main points that should be taken into account in evaluating (3.2). Some extensions to the basic procedure are also surveyed. In Section 3.3 the calculation of expected values is described, and the use of uniformization to calculate steady state measures is briefly discussed. In Section 3.4 the problem of calculating probability distributions is addressed. We survey a technique used to calculate the distribution of accumulated reward based on both uniformization and the Laplace transform. Then we describe a methodology based exclusively on the probabilistic properties of uniformization. Section 3.5 concludes the chapter.

3.2 The uniformization technique

3.2.1 The basic approach

We consider a homogeneous continuous-time Markov chain $\mathcal{X} = \{X(t) : t \geq 0\}$ with finite state space $\mathcal{S} = \{1, \ldots, M\}$ and infinitesimal generator

$$\mathbf{Q} = \begin{bmatrix} -q_1 & q_{12} & \cdots & q_{1M} \\ q_{21} & -q_2 & \cdots & q_{2M} \\ \vdots & \vdots & \ddots & \vdots \\ q_{M1} & q_{M2} & \cdots & -q_M \end{bmatrix}.$$

For $i \neq j$, entry q_{ij} of $\mathbf{Q}$ is the exponential rate at which transitions of the chain occur from i to j, while $q_i = \sum_{j \neq i} q_{ij}$ is the exponential rate out of state i.

Let $\mathbf{P}(t)$ be the matrix for which the (i,j)th element is the probability that the chain is in state j at time t starting from state i. Also recall that $\mathbf{p}(t)$ is the vector whose ith element is the probability that the chain is in state i at time t given an initial distribution.

For the Markov chain $\mathcal{X}$, the probability p_{ij} of moving to state j when a transition from i occurs is $q_{ij}/\sum_{j \neq i} q_{ij}$, and the residence time spent in a visit to i is exponential with rate q_i. Assume that we transform the chain by allowing transitions back to the same state, so that each time state i is visited the process spends an exponential amount of time there with rate $\Lambda_i \geq q_i$ before either leaving to another state or immediately returning to i. Furthermore, we preserve the probability of moving from i to j given that the transition is to a state other than i. Let us call this new process $\mathcal{X}_i^*$. Let p_{ij}^* be the probability that a transition of $\mathcal{X}_i^*$ from i to j occurs. Then $p_{ij}^* = q_{ij}/\Lambda_i$, while $p_{ii}^* = 1 - q_i/\Lambda_i$.

In $\mathcal{X}_i^*$, the number of visits to i before exiting has a geometric distribution with parameter p_{ii}^*, and the total amount of time spent in i before moving to a different state is exponential with the same rate q_i as in the original process $\mathcal{X}$. Since the total residence time in i is identical in both processes, as is the probability of moving from i to j given that a transition occurs to a state other than i, we may consider $\mathcal{X}$ and $\mathcal{X}_i^*$ as equivalent processes.

Now choose $\Lambda = \max_i\{\Lambda_i\}$, and perform this transformation on every state to yield a process $\mathcal{X}^*$. Then we have uniformized the exponential rates which govern the time until a transition occurs in $\mathcal{X}^*$, each such transition being either to the same state or to a different one. As a consequence, the times of the transitions are given by a Poisson process of rate Λ, and the transition probabilities are given by a discrete-time Markov chain. In more detail, the number of transitions in an interval $(0, t)$ is given by a Poisson process with rate Λ, and $\mathcal{X}^*$ can be considered to be a Markov chain subordinated to a Poisson process as follows. Let $\mathcal{Z} = \{Z_n : n = 0, 1, \ldots\}$ be a discrete-time Markov chain with finite state space $\mathcal{S}$ and one-step transition probability matrix $\mathbf{P} = \mathbf{I} + \mathbf{Q}/\Lambda$, and let $\mathcal{N} = \{N(t) : t \geq 0\}$ be a Poisson process with rate Λ which is independent of $\mathcal{Z}$. Then, we may interpret $X(t) = Z_{N(t)}$ for $t \geq 0$.

After n transitions, $\mathcal{X}^*$ is in state j with probability $v_j(n)$, where $v_j(n)$ is the jth entry of the vector $\mathbf{v}(n) = \mathbf{v}(0)\mathbf{P}^n$ and $\mathbf{v}(0)$ is the initial state probability vector. Unconditioning on the number of transitions in $(0, t)$, we obtain

$$\mathbf{p}(t) = \sum_{n=0}^{\infty} e^{-\Lambda t} \frac{(\Lambda t)^n}{n!} \mathbf{v}(n), \tag{3.3}$$

which is the basic equation of the uniformization method as stated in the introduction.

Equation (3.3) has many interesting properties. First, only non-negative elements are involved in the computation of (3.3), which helps to avoid numerical problems. Also, $\mathbf{v}(n)$ can be evaluated recursively by $\mathbf{v}(n) = \mathbf{v}(n-1)\mathbf{P}$ with $\mathbf{v}(0) = \mathbf{p}(0)$, and so if $\mathbf{P}$ is sparse, which is true for matrices of many models, considerable savings are obtained. Furthermore, as shown below, it is easy to obtain error bounds when the infinite series in (3.3) is truncated. As mentioned in the introduction, one of the main advantages of the technique is its probabilistic interpretation. The state transition times are governed by a Poisson process *independent* of $\mathcal{Z}$, and many properties of the Poisson process can be used to obtain more complex performability measures than $\mathbf{p}(t)$ as given in (3.3). One such useful property is the *exchangeability* of the interval lengths between transition times, which is discussed in Section 3.4.

Several considerations must be made in computing with (3.3). First, the infinite series must be truncated. However, error bounds can be easily calculated from the properties of the Poisson distribution as follows. If we truncate (3.3) at N and note that $v_j(n) \le 1$, then the error $\varepsilon(N)$ of any entry in the vector $\mathbf{p}(t)$ is clearly given by

$$\varepsilon(N) \le 1 - \sum_{n=0}^{N} e^{-\Lambda t}\frac{(\Lambda t)^n}{n!}. \tag{3.4}$$

As a consequence, N may be chosen before starting the calculations for a given error tolerance. For example, if $\varepsilon(N) < 10^{-3}$ and $\Lambda t = 1$, then $N = 5$; if $\Lambda t = 10$, then $N = 21$ for the same error tolerance. For large Λt, the Poisson distribution can be approximated by the normal distribution with parameters $(\Lambda t, \Lambda t)$ (see [189]). If Ω is a normal $(0, 1)$ random variable and z_ε is such that $\Pr\{\Omega > z_\varepsilon\} \le \varepsilon$, then

$$N \approx \Lambda t + z_\varepsilon\sqrt{\Lambda t}.$$

For instance, if $\Lambda t = 100$ and $\varepsilon(N) < 10^{-3}$, then $N \approx 131$. As discussed in the following sections, the calculation of many performability measures requires the truncation of an infinite series, and the truncation error can be bounded in a similar manner.

Calculating the Poisson distribution in (3.3) in a straightforward manner may cause overflow/underflow problems, even for moderate values of Λt. Let $\beta(n) = e^{-\Lambda t}(\Lambda t)^n/n!$. Then an integer $N_{\min}$ can be chosen such that the lower tail of the Poisson distribution is $\varepsilon' \ll \varepsilon$ but greater than the underflow limit. From the normal approximation of the Poisson distribution, we have

$$N_{\min} \approx \max(0, \Lambda t - z_{\varepsilon'}\sqrt{\Lambda t}).$$

This may not be sufficient to avoid underflow when calculating $\beta(N_{\min})$ if $N_{\min}$ is positive. However, $\beta(N_{\min})$ can be calculated from

$$\beta(N_{\min}) = e^{-\Lambda t/(N_{\min}+1)} \prod_{i=1}^{N_{\min}} \frac{E}{i},$$

where $E = \Lambda t e^{-\Lambda t/(N_{\min}+1)}$. The product is evaluated as follows. The initial value of i is chosen such that $E/i \approx 1$, and then one of the indices $j = i+1$ or $j = i-1$ is chosen in order to maintain the intermediate result as close to 1 as possible. It can be shown that this procedure is guaranteed to yield $\beta(N_{\min})$ without underflow/overflow problems (see [106]).

Many models have a sparse generator $\mathbf{Q}$, and consequently $\mathbf{P}$ is also sparse. The recursion $\mathbf{v}(n) = \mathbf{v}(n-1)\mathbf{P}$ is a simple vector matrix multiplication, and this operation should exploit the sparseness of $\mathbf{P}$.

The number of operations to compute (3.3) depends on the value of N and the sparseness of $\mathbf{P}$. If d is the average number of transitions out of a state, i.e., the average number of nonzero off-diagonal entries in a row of $\mathbf{P}$, then the number of operations is $O(NMd)$. The value of N depends on Λt, while Λ depends on the largest total output rate q_i among the states.

In many applications one is interested in computing $\mathbf{p}(t)$ for several time points, say for instance, $\tau_1 < \tau_2 < \cdots < \tau_L$. The computational effort to compute $\mathbf{p}(\tau_i)$, $i = 1, \ldots, L$, is practically the same as the effort to compute $\mathbf{p}(\tau_L)$. From (3.3) we have (for $i = 1, \ldots, L$)

$$\mathbf{p}(\tau_i) = \sum_{n=0}^{\infty} e^{-\Lambda\tau_i} \frac{(\Lambda\tau_i)^n}{n!} \mathbf{v}(n).$$

The infinite series for τ_L must be truncated at a limit N so that the error ε satisfies the bound desired for τ_L, and consequently the bound for each $\tau_i < \tau_L$. The most expensive part of using the above equation for computing the $\mathbf{p}(\tau_i)$ is in the calculation of $\mathbf{v}(n)$, which does not depend on the value of τ_i. It only once needs to be computed for all τ_i, $i = 1, \ldots, L$.

Thus far we have considered the finite state space case. However, models such as queueing systems with infinite capacity have an infinite state space. One approach for handling this type of model is to truncate the state space, say to M states, and aggregate all other states into a single absorbing state. It is easy to see that the additional error introduced with this approximation is bounded by the probability that the system is in this absorbing state (see also [123]).

3.2.2 Extensions to the basic approach

Various extensions can be made to the basic uniformization method, either to improve the computational effort in evaluating (3.3) or to expand the applicability of the method. We now discuss several of these extensions.

The nonhomogeneous case

The uniformization technique may be applied to nonhomogeneous Markov chains for which changes in the infinitesimal generator $\mathbf{Q}$ occur only at a finite number of time points $\tau_1, \tau_2, \ldots, \tau_K$, i.e., all rates are constant during the fixed intervals (τ_{i-1}, τ_i), $i = 1, \ldots, K$ (here $\tau_0 = 0$ by definition). Let $\mathbf{P}_i$ be the uniformized transition probability matrix corresponding to the interval (τ_{i-1}, τ_i), and let $\mathbf{p}(\tau_i)$ be the state probability vector at time τ_i. Then

$$\mathbf{p}(\tau_i) = \sum_{n=0}^{\infty} e^{-\Lambda \delta_i} \frac{(\Lambda \delta_i)^n}{n!} \mathbf{v}_i(n), \tag{3.5}$$

where $\delta_i = \tau_i - \tau_{i-1}$ and $\mathbf{v}_i(n) = \mathbf{v}_i(n-1)\mathbf{P}_i$. Starting from $i = 1$, the state probabilities are calculated at each given time point and then used as the initial state probabilities for the next interval, i.e., $\mathbf{v}_i(0) = \mathbf{p}(\tau_{i-1})$.

When applying equation (3.5), care must be taken in choosing the truncation values N_i to obtain the desired error tolerance. It is easy to show that the truncation errors $\varepsilon_i = \varepsilon_i(N_i)$, $i = 1, \ldots, K$, accumulate from one interval to the next. Thus the values of N_i should be chosen so that all the errors ε_i (for $i = 1, \ldots, K$) are proportional to the length of the corresponding interval and so that their sum is at most the total given error tolerance.

Nonhomogeneous Markov chains with time-dependent infinitesimal generator $\mathbf{Q}_t$ have been considered in [124]. Let $\mathbf{P}_t$ be the corresponding uniformized one-step transition probability matrix for time t. If we condition on n transitions by time t and on the times $\tau_1, \ldots, \tau_n$ of these transitions, then $\mathbf{P}(t)|n$, the conditional state probability matrix at time t, is simply $\mathbf{P}_{\tau_1}\mathbf{P}_{\tau_2}\cdots\mathbf{P}_{\tau_n}$. The number of transitions during $(0,t)$ has a Poisson distribution, and the joint distribution of $\tau_1,\ldots,\tau_n$ is the same as the distribution of the order statistics of a set of n independent and identically distributed random variables uniform on $(0,t)$, so $\mathbf{P}(t)$ can be obtained by unconditioning using the known joint density of $\tau_1,\ldots,\tau_n$. The resulting expression involves a multiple integral, and in order to carry out computations, time is discretized and the integrals are approximated by finite sums. Although the approach is of theoretical interest, it is very costly from a computational point of view.

Dynamic generation

It may be computationally advantageous to generate the uniformized matrix $\mathbf{P}$ dynamically [194]. In other words, we generate $\mathbf{P}$ in parallel with the computation of $\mathbf{v}(n)$. This is possible if we observe that, at step n in (3.3), only states reachable in n steps from an initial set need to be considered.

Let $\mathcal{A}(n)$ be the set of *active* states at step n, i.e., the set of states that have nonzero probability at step n. If the cardinality of $\mathcal{A}(0)$ is small and if each state does not have many outgoing transitions, then $\mathcal{A}(n)$ may have small cardinality in comparison with that of the entire state space. For instance, in an M/M/1-type queueing process with initial state 0, the cardinality of $\mathcal{A}(n+1)$ increases by only 1 over that of $\mathcal{A}(n)$, and $\mathbf{P}^{(n+1)}$ (the matrix generated at step $n+1$) has one more row and column than $\mathbf{P}^{(n)}$. Note that only the active states at step n need to be considered in the calculation of $\mathbf{v}(n+1)$. This procedure is also useful for infinite state space models, provided that the transition rates as well as the number of transitions out of any state are bounded.

Models with large Λt

The computational complexity of the uniformization technique in calculating $\mathbf{p}(t)$ is proportional to Λt, where $\Lambda \geq \max_i\{q_i\}$. In other words, the number of operations necessary to compute $\mathbf{p}(t)$ increases with the product of t, the observation period length, and $\max_i\{q_i\}$, the largest value of the total output rates from the states. A model is called *stiff* if it contains output rates with both large and small values. Such a difference in rates gives rise to large Λt values, and thus stiff models require substantial computation to be solved.

Several methods have been proposed to improve the efficiency of the uniformization technique for models with large Λt values. In one method, called selective randomization [311], the state space is divided into two disjoint sets. The largest value of q_i is obtained for both sets, and uniformization is selectively applied to each set. The main problem with this procedure is that it requires maintaining a counter of the number of transitions in one of the subsets after a total of n transitions, and this in general substantially increases the state space of the original process.

Another method, called adaptive uniformization [332], is in the same spirit as the dynamic generation method described above. In the dynamic generation method, it is assumed that the value of Λ is known, i.e., the largest total output rate is known. However, in adaptive uniformization, a possible new value, say $\Lambda(n)$, is computed at each step n according to the corresponding active states. The idea is that using $\Lambda(n)$, which always satisfies $\Lambda(n) \leq \Lambda$, may possibly lead to computational savings (note that Λ need not be known in advance). Let $\mathbf{P}^{(n)}$ be the uniformized matrix corresponding to the set of active states after the nth transition takes place. If the state probabilities $\mathbf{v}(n)$ at the end of the nth transition are known, then after the $(n+1)$th transition we have

$$\mathbf{v}(n+1) = \mathbf{v}(n)\mathbf{P}^{(n)}.$$

Each transition now occurs at a possibly different rate. Let $W_n(t)$ be the probability that n transitions occurred by time t. Then

$$\mathbf{p}(t) = \sum_{n=0}^{\infty} W_n(t)\mathbf{v}(n). \tag{3.6}$$

It remains to compute $W_n(t)$, the probability of exactly n transitions by time t of a pure birth process with possibly different rates for each state. In [332] the issues involved in this calculation and ways to efficiently obtain $W_n(t)$ are discussed, while in [133] different methods of calculating it are compared.

Bobbio and Trivedi [37] used an aggregation technique to deal with stiff problems, and this method has been used in conjunction with uniformization [387]. In their approach the states of the model are partitioned into two sets, such that one of the sets contains only states with small output rates (called slow states), while the complement includes all states with at least one large output rate (the fast states). States in the fast subset are further divided into "nearly transient" and "nearly recurrent" states according to the graph connecting these states with the slow rate transitions removed. The key assumption used is that the subset of fast nearly recurrent states reaches steady state by time t. The subset containing fast nearly recurrent states is then aggregated, and the transient solution for the slow states is found from the aggregated model. Fast nearly transient states are eliminated by known techniques. The final solution is obtained after a disaggregation step.

Abdallah and Marie [1] have proposed a technique called the uniformized power method. For a time τ, recall that $\mathbf{P}(\tau)$ is the matrix whose (i,j)th element is equal to the probability that the continuous-time Markov chain $\mathcal{X}$ is in state j at time τ starting from state i. Suppose that the observation period length t is expressed in the form $t = t_n = 2^n t_0$ for some n and some t_0. If t_0 is chosen such that Λt_0 is small, then $\mathbf{P}(t_0)$ can be calculated using uniformization with small computational effort. From the Chapman-Kolmogorov equations, $\mathbf{P}(t_k) = \mathbf{P}(t_{k-1})^2$, and so $\mathbf{P}(t) = \mathbf{P}(t_n)$ can be obtained by performing n matrix multiplications after $\mathbf{P}(t_0)$ is calculated. Finally, the transient solution is obtained from $\mathbf{p}(t) = \mathbf{p}(0)\mathbf{P}(t)$.

Although $\mathbf{P}(t_0)$ is obtained within a given error tolerance, care must be taken to evaluate the final error of the method after the n matrix products are performed. In [1], a new approach was proposed to bound the final error. For large Λt values, the method of [1] can be used to compute the desired result in a much smaller number of steps than regular uniformization. But note that each step requires multiplication of matrices which are not sparse. This contrasts with standard uniformization, which does take advantage of and preserves the sparseness of $\mathbf{P}$, a property that is common in many problems, and involves only vector matrix multiplications.

An approach, called steady-state detection, that is useful in some problems with large Λt values was proposed in [77]. Consider equation (3.3) and recall that $\mathbf{v}(n) = \mathbf{v}(0)\mathbf{P}^n$. If $\mathbf{v}(n)$ approaches its steady-state limit for $n = K < N$, where N is the truncation point of the infinite series in (3.3), then $\mathbf{v}(K) \approx \mathbf{v}(K+1)$ and (3.3) can be approximated by

$$\mathbf{p}(t) \approx \sum_{n=0}^{K-1} e^{-\Lambda t}\frac{(\Lambda t)^n}{n!}\mathbf{v}(n) + \left[1 - \sum_{n=0}^{K-1} e^{-\Lambda t}\frac{(\Lambda t)^n}{n!}\right]\mathbf{v}(K). \qquad (3.7)$$

If $K \ll N$ (recall that $N \approx \Lambda t$ for large Λt), then significant savings can be realized. Note that, from (3.7), the number of steps to obtain the solution is equal to the number of steps until convergence is achieved if the power method is used to obtain the steady state solution. Therefore, this number depends on the subdominant eigenvalue of the transition matrix $\mathbf{P}$ and not on Λt. One problem with the method, however, is the detection of convergence to steady state, and care must be taken when convergence of the power method is slow. If the steady state model solution is available, it can be used to detect whether $\mathbf{v}(n)$ has reached steady state. Note also that, unlike standard uniformization, there is no ability to specify error bounds in advance.

However, if the measure of interest is the probability that the model is in a subset of states by time t and not the individual state probabilities at t, one can stop the computation at any step and calculate error bounds as shown in [422]. Let $\boldsymbol{\sigma}(n) = \mathbf{P}^n \mathbf{1}_{\mathcal{U}}$, where the ith entry of $\mathbf{1}_{\mathcal{U}}$ is equal to 1 if state i is in subset $\mathcal{U} \subset \mathcal{S}$ and is 0 otherwise. Then $\sigma_i(n)$, the ith entry of $\boldsymbol{\sigma}(n)$, is the probability that the chain is in subset $\mathcal{U}$ after the nth transition given that it started in state i. Define $m_n = \min_{i \in \mathcal{S}} \{ \sigma_i(n) \}$ and $M_n = \max_{i \in \mathcal{S}} \{ \sigma_i(n) \}$. Since $\sigma_i(n+1) = \sum_{j=1}^{M} p_{ij} \sigma_j(n)$, it is easy to see that $m_n \leq \sigma_i(n+1) \leq M_n$ for all i, and the sequences m_n and M_n are non-decreasing and non-increasing, respectively. If steady state is reached at step L, then $M_L \approx m_L$ and

$$\Pr\{X(t) \in \mathcal{U}\} \approx \sum_{n=0}^{L} e^{-\Lambda t} \frac{(\Lambda t)^n}{n!} \boldsymbol{\pi}(0) \boldsymbol{\sigma}(n) + m_L \left[1 - \sum_{n=0}^{L} e^{-\Lambda t} \frac{(\Lambda t)^n}{n!} \right],$$

with the error bound being obtained from $|M_L - m_L|$.

Recently, a new method called regenerative randomization [59] was proposed to handle models for which the product Λt is very large. Let $\mathcal{X}$ be the original continuous-time Markov chain, and let $\mathcal{Z}$ be the underlying discrete-time Markov chain obtained after uniformization. A state, say u, of $\mathcal{X}$ (equivalently, a state of $\mathcal{Z}$) is selected as a *regeneration state*. In order to simplify the presentation, assume that the initial state of the chain $\mathcal{X}$ (equivalently, $\mathcal{Z}$) is u. A new discrete-time Markov chain, say $\mathcal{W}$, is defined with states which count the number of steps $\mathcal{Z}$ makes between visits to state u. In other words, state i of $\mathcal{W}$ indicates that $\mathcal{Z}$ has made i transitions since its last visit to u, while state 0 represents a visit to u in $\mathcal{Z}$. From state i of $\mathcal{W}$ only states $i+1$ or 0 can be visited. The transition probabilities of W can be easily obtained from an auxiliary transient chain $\mathcal{U}$, which is identical to $\mathcal{Z}$ except that it has u as an absorbing state.

Let W^* be the continuous-time chain with the same state space as W, for which the rates out of any state of W^* are equal to Λ and the transition probabilities of going from i to j are identical to those for chain W. In [59] it is shown that $\mathbf{p}(t)$, the probability distribution for $X(t)$, can be obtained from W^* and $\mathcal{U}$. Although the chain W^* is stiff, its simple structure allows it to be solved explicitly via Laplace transforms. Note that, if the chain W^* is in state n at time t, then the probability of $\mathcal{X}$ being in state l by that time is the probability that $\mathcal{U}$ is in state l after n steps, given that it has not reached the absorbing state in n steps. Intuitively, if W visits state u frequently, then the probability of making a large number of transitions between visits of u is small. In that case, we only need to calculate the state probabilities for a small number of states of $\mathcal{U}$ (i.e., a small number of steps away from u) to obtain $\mathbf{p}(t)$ within the desired error tolerance. In [59] several examples are given which illustrate the efficacy of the approach in comparison with standard uniformization. Clearly, the choice of u affects the efficiency of the method.

3.3   Expected values

In this section we indicate how the uniformization technique can be used to calculate important performability measures of Markov reward models involving the expected value of various random variables. In performability modelling (see, for instance, [81, 106, 315, 316, 319] and the references therein), rewards are associated with states or with transitions (pairs of states) of the model. Rewards associated with a state are reward rates gained by the system per unit time in the corresponding states. One example is the operational time of a system during an interval, for which a reward rate of 1 per time unit is associated to the operational states of the model and a reward of 0 is associated to the failed states. The total reward accumulated during the interval is the total time for which the system remains in the operational states. Rewards associated with transitions are impulse rewards gained each time an event of interest takes place. For instance, the modeller may be interested in counting the number of times events of a particular type occur. As an example, suppose our interest is in the number of times the system failed due to the failure of a given component. In this case, an impulse reward of 1 is associated with transitions that correspond to the given component failing and leading to a state representing a nonoperational system.

3.3.1 Basic results

Let us assume that there are $K + 1$ distinct reward rates $r_1^* > r_2^* > \cdots > r_{K+1}^*$, where without loss of generality $r_1^* = 1$ and $r_{K+1}^* = 0$. Let r_i be the reward associated with state i, and so r_i can take any of the $K + 1$ values $r_l^*, l = 1, \ldots, K + 1$. The vector of reward rates associated with the states is denoted by $\mathbf{r} = \langle r_1, \ldots, r_M \rangle$. Similarly, we assume that there are $\widehat{K} + 1$ impulse rewards $\widehat{r}_1^* > \widehat{r}_2^* > \cdots > \widehat{r}_{\widehat{K}+1}^*$, with $\widehat{r}_1^* = 1$ and $\widehat{r}_{\widehat{K}+1}^* = 0$. Let $\widehat{r}_{ij}$ be the reward associated with the transition from i to j, i.e., with the pair of states (i, j). The matrix of impulse rewards associated with transitions is denoted by $\widehat{\mathbf{R}} = [\widehat{r}_{ij}]$. We also introduce the following notation. Given two non-negative vectors $\mathbf{x} = \langle x_1, \ldots, x_M \rangle$ and $\mathbf{y} = \langle y_1, \ldots, y_M \rangle$, then their inner product is denoted by $\mathbf{x} \cdot \mathbf{y} = \sum_{i=1}^M x_i y_i$. Also, the (sum) norm of $\mathbf{x}$ is denoted by $\|\mathbf{x}\| = \sum_{i=1}^M x_i$.

Consider the case of models with reward rates. The random variable $r_{X(t)}$ is the instantaneous reward at time t. We define the point performability $PP(t)$ as the expected reward at time t, namely,

$$PP(t) = E[r_{X(t)}] = \mathbf{r} \cdot \mathbf{p}(t).$$

Then from (3.3),

$$PP(t) = \sum_{n=0}^{\infty} e^{-\Lambda t} \frac{(\Lambda t)^n}{n!} \{\mathbf{r} \cdot \mathbf{v}(n)\}. \tag{3.8}$$

Let $Y(t)$ be the reward accumulated in $(0, t)$. As an example, assume that the structure of a system is modelled using a Markov chain, for which each state represents a given system configuration, and associate a reward to each state equal to the system throughput (jobs completed per unit of time). Then $Y(t)$ is the total number of jobs completed by time t. The cumulative reward averaged over the observation period is $Y(t)/t$. The time to achieve a given reward level r, $\Theta(r)$, is immediately obtainable from $Y(t)$, since the event $Y(t) > r$ is equivalent to the event $\Theta(r) < t$ for non-negative rewards.

The random variable $Y(t)$ may be found by integrating the instantaneous reward $r_{X(t)}$ as

$$Y(t) = \int_0^t r_{X(s)} ds.$$

Then, from (3.8), the expected value of $Y(t)$ is

$$\begin{aligned}
E[Y(t)] &= \int_0^t E[r_{X(s)}] ds \\
&= \int_0^t \sum_{n=0}^{\infty} e^{-\Lambda s} \frac{(\Lambda s)^n}{n!} \{\mathbf{r} \cdot \mathbf{v}(n)\} ds \\
&= \frac{1}{\Lambda} \sum_{n=0}^{\infty} \{\mathbf{r} \cdot \mathbf{v}(n)\} \int_0^t e^{-\Lambda s} \frac{\Lambda(\Lambda s)^n}{n!} ds.
\end{aligned}$$

Noting that the term inside the integral is the $(n+1)$-stage Erlangian density, we finally have

$$E[Y(t)] = \frac{1}{\Lambda} \sum_{n=0}^{\infty} E_{n+1,\Lambda}(t)\{\mathbf{r} \cdot \mathbf{v}(n)\}, \tag{3.9}$$

where $E_{n+1,\Lambda}(t) = 1 - \sum_{j=0}^{n} e^{-\Lambda t} \frac{(\Lambda t)^j}{j!}$ is the $(n+1)$-stage Erlangian distribution.

Substituting the formula for $E_{n+1,\Lambda}(t)$ into (3.9) and exchanging the order of summation, we have the alternative expression

$$E[Y(t)] = t \sum_{n=0}^{\infty} e^{-\Lambda t} \frac{(\Lambda t)^n}{n!} \left[\frac{\sum_{j=0}^{n} \{\mathbf{r} \cdot \mathbf{v}(j)\}}{n+1} \right]. \tag{3.10}$$

The term inside the brackets can be evaluated recursively as

$$f(n) = \frac{n}{n+1} f(n-1) + \frac{1}{n+1} \{\mathbf{r} \cdot \mathbf{v}(n)\},$$

where $f(0) = \mathbf{r} \cdot \mathbf{v}(0)$.

Equation (3.10) can also be derived by simple probabilistic arguments. Given n transitions of the uniformized process during $(0, t)$, the observation period is divided into $n + 1$ intervals, which are determined by the associated independent Poisson process $\mathcal{N}$. It is well known that the expected length of each subinterval is $t/(n+1)$ from results for the Poisson process (see, for instance, [390]). Since the probability distribution associated with the jth interval is $\mathbf{v}(j)$, the product of t and the term in brackets in (3.10) is the expected accumulated reward given n transitions.

Similarly to the calculation of $\mathbf{p}(t)$, an infinite sum must be truncated in order to evaluate equation (3.9) or equation (3.10) above. Since the term in brackets in (3.10) is upper bounded by 1 (recall that the rewards were assumed to be between 0 and 1), then clearly the truncation error in (3.10) is bounded by t times that in (3.4). In [192] it was noted that

$$\frac{1}{\Lambda t} \sum_{n=N+1}^{\infty} E_{n+1,\Lambda}(t) < \sum_{n=N+1}^{\infty} e^{-\Lambda t} \frac{(\Lambda t)^n}{n!},$$

and so (3.9) can be truncated at a smaller value of N than (3.10) in order to obtain the same error tolerance.

We now consider models with impulse rewards. Let $NR(t)$ be the accumulated reward obtained from transitions during a finite interval $(0, t)$, and we wish to find its expected value. The $M \times M$ matrix that gives the reward earned after one step of the discrete-time Markov chain $\mathcal{Z}$ has its (i, j) entry given by $\hat{r}_{ij} p_{ij}$, i.e., this matrix is $\widehat{\mathbf{R}} \circ \mathbf{P}$, the Hadamard product (see [234]) of the reward matrix $\widehat{\mathbf{R}}$ and the transition matrix $\mathbf{P}$. Conditioned on n transitions of $\mathcal{Z}$, then the accumulated reward is the sum of the rewards earned in each transition. If the state probability vector at the end of the $(k-1)$th transition is known, the expected value of the reward earned in the kth transition is simply $\|\mathbf{v}(k-1)(\widehat{\mathbf{R}} \circ \mathbf{P})\|$. Summing the rewards over all n transitions and unconditioning on the number of transitions, we have

$$E[NR(t)] = \sum_{n=1}^{\infty} e^{-\Lambda t} \frac{(\Lambda t)^n}{n!} \sum_{k=1}^{n} \|\mathbf{v}(k-1)(\widehat{\mathbf{R}} \circ \mathbf{P})\|. \tag{3.11}$$

3.3.2 Additional results

The uniformization technique has also been used as an intermediate step as part of a model solution. Useful examples for performability modelling include the calculation of the entries of transition rate matrices, and the calculation of transient measures necessary during the solution process to obtain steady-state measures for some models. We briefly survey some basic approaches that can also be applied in a more general setting than performability modelling.

Grassmann [191] used uniformization to calculate the transition probability matrix for the GI/PH/1 queue. The matrix entries are calculated from an embedded Markov chain at instants immediately before arrivals. If the time between arrivals is constant, then the probability of a transition from one state to another can be obtained by using uniformization on the Markovian departure process. If the interarrival times are not constant, one has to obtain the state probabilities by conditioning on the number of steps of the uniformized departure process between embedded points. Several interarrival time distributions are considered in [191].

In [105] uniformization was used to analyze scheduled maintenance policies of computer systems. The times between scheduled maintenances are not necessarily constant, and the system is allowed to undergo unscheduled repairs between scheduled maintenances. Embedded points were identified at scheduled maintenance times, and an embedded Markov chain was obtained. Its transition probabilities were determined from the *transient* solution of a continuous-time Markov chain that describes the failure and repair behaviour of the system between arrivals of the scheduled repair person. Measures such as the system availability can be calculated from the solution of the embedded chain. Several different repair policies were considered in [105]. For example, in one of the policies, if the system breaks down and an unscheduled repair takes place, the next scheduled maintenance is rescheduled after the repair is completed. In another policy, the next scheduled maintenance is not rescheduled.

There is a large class of non-Markovian models for which embedded points can be identified, and which requires the use of transient analysis to find both the transition probabilities between embedded points and the performance/dependability measures of interest, once the state probabilities of the embedded chain are calculated. As in the examples given above, uniformization has been applied to these two steps of the model solution. In [290] uniformization was employed in obtaining the transition probabilities of a subset of Deterministic and Stochastic Petri-Net (DSPN) models. In [112] a methodology was developed which addresses the calculation of transient probabilities and associated measures of interest for a broad class of models, and which takes advantage of special model structure to reduce the computational complexity of the solution. One of the main advantages of the technique is that probabilistic reasoning can be applied to aid in the development of the solution. Recently in [161], an iterative method was devised to avoid the storage of the embedded matrix calculated during the intermediate step of the model solution. The approach can be applied to models with transition distributions of the form considered in [191].

Traffic characterization and modelling has been an extensive area of research, and in the last several years many studies have been performed to obtain accurate models for network resource dimensioning. Several of these are Markov reward models (fluid-flow models) for which a reward rate is associated with each state representing, for instance, the traffic flow sent by the source. In general the models are designed so that first- and second-order statistics (traffic descriptors) from real traffic traces are matched against those obtained from the model. These descriptors include, for example, the mean traffic rate, the burstiness, the autocovariance $Cov(\tau) = E[X(t), X(t + \tau)] - E[X(t)]^2$, and the index of dispersion for counts $IDC(\tau) = Var[N(t)]/E[N(t)]$ for a given time lag τ, where $X(t)$ is the traffic rate at t and $N(t)$ is the process that counts the amount of traffic transmitted during $(0, t)$. These last two descriptors capture traffic correlations and the variability of the amount of traffic transmitted in an interval. In [276], recursions were obtained to calculate $Cov(\tau)$ and $IDC(\tau)$. The calculation of these descriptors requires the use of transient analysis over a given time interval, and the uniformization technique was the basis of the method (see [276] for details).

3.4 Probability distributions

We now discuss the use of the uniformization technique to obtain distributions of several random variables of interest. Specifically, we consider in more detail the distribution of accumulated reward during a finite observation period. We start by describing a recent method proposed in the literature that is based on both uniformization and the Laplace transform. Then we present the methodology developed in [104] for calculating transient performability measures, which is based on the probabilistic properties of uniformization. We also comment on new results that apply the methodology to efficiently calculate distributions of accumulated reward.

Many Laplace transform approaches for calculating performability measures are based on recursive expressions obtained after conditioning on the time of the first transition of a continuous-time Markov chain. Let

$F_i(y,t) = \Pr\{Y(t) \le y | X(0) = i\}$. In [135] the Markov chain $\mathcal{X}$ is first uniformized, and so the time between transitions is given by an exponential distribution with rate Λ. Conditioning on the number of transitions during $(0, t)$, we have

$$F_i(y,t) = \sum_{n=0}^{\infty} e^{-\Lambda t} \frac{(\Lambda t)^n}{n!} F_i(y, t|n). \tag{3.12}$$

Given n transitions in $(0, t)$, it is well known that the time of the first transition, say τ_1, has density function (for $\tau < t$)

$$\frac{d\Pr\{\tau_1 \le \tau\}}{d\tau} = \frac{n}{t}\left[1 - \frac{\tau}{t}\right]^{n-1}.$$

Therefore, conditioning on the time of the first transition, we obtain

$$F_i(y,t|n) = \int_0^t \frac{n}{t}\left[1 - \frac{\tau}{t}\right]^{n-1} \sum_{j=1}^{M} p_{ij} F_j(y - r_i\tau, t - \tau|n-1)\,d\tau \tag{3.13}$$

Taking the double Laplace transform of equation (3.13) with respect to the y and t variables, performing a partial fraction expansion and inverting the result, an expression for $F_i(y,t|n)$ is found in terms of coefficients that can be calculated recursively (see [135] for more details).

In what follows we present a methodology based on the probabilistic properties of uniformization. Recall that in calculating the performability measures of interest reward rates are assigned to states and impulse rewards are assigned to transitions. We colour the $n+1$ intervals according to the reward rate associated with the corresponding state of chain $\mathcal{Z}$, or according to the impulse reward associated with transitions at the beginning or the end of the interval (i.e., with pairs of states of $\mathcal{Z}$). For instance, if we are interested in the system availability, we colour the intervals corresponding to an operational system according to reward rate 1 and the remaining states according to reward rate 0. In the general performability case, there are $K+1$ distinct colours corresponding to reward rates. If the system is in state j during the ith interval and the reward rates are associated with the states, we colour interval i according to r_j, the reward assigned to state j. The measures to be calculated are obtained as functions of the number or length of the intervals with the same colour.

Consider a model with reward rates, and let $M(t)$ be the measure of interest. Given that n transitions have occurred during the observation period, the interval $(0, t)$ is divided into $n+1$ subintervals by these events. Associated with each subinterval is the corresponding state of the uniformized Markov chain $\mathcal{Z}$ and its reward rate. We assign colour l to a subinterval if the associated state has reward rate r_l^*. Assume that there are k_l intervals with colour l for $l = 1, \ldots, K + 1$. Define $\mathbf{k} =< k_1, \ldots, k_{K+1} >$, i.e., $\mathbf{k}$ is the vector that gives the number of intervals corresponding to a given colour. We refer to a specific vector $\mathbf{k}$ as a (rate) colouring. Let $\Gamma[n, \mathbf{k}]$ be the probability of the colouring $\mathbf{k}$ given n transitions for measure $M(t)$. Then

$$M(t) = \sum_{n=0}^{\infty} e^{-\Lambda t} \frac{(\Lambda t)^n}{n!} \sum_{||\mathbf{k}||=n+1} \Gamma[n, \mathbf{k}] M(t, n, \mathbf{k}), \qquad (3.14)$$

where $M(t, n, \mathbf{k}) = M(t)$, *given n transitions and colouring $\mathbf{k}$*. The goal is to find (calculate) $\Gamma[n, \mathbf{k}]$ and $M(t, n, \mathbf{k})$ in an efficient manner.

Similarly, for a model with impulse rewards, we define an impulse colouring as a vector $\widehat{\mathbf{k}} = \langle \widehat{k}_1, \ldots, \widehat{k}_{\widehat{K}+1} \rangle$ the ith entry of which represents the number of transitions with impulse reward $\widehat{r}_i^*$. Given an impulse measure $\widehat{M}(t)$, an equation similar to (3.14) can be derived.

We first consider an impulse measure to illustrate the methodology. Let us assume that our interest is in the distribution of the number of events of a certain type (for example, the number of times the system failed during the observation period). In this case there are only two impulse rewards (0 and 1), and colouring the intervals reduces to marking those that are entered via a particular type of transition. Let $\Upsilon(t)$ be the random variable that gives the number of events of interest during $(0, t)$, and suppose that the measure to be calculated is the distribution function of $\Upsilon(t)$, i.e., $\widehat{M}(t) = \Pr\{\Upsilon(t) \leq \widehat{r}\}$. In this case, an impulse colouring vector $\mathbf{k}$ has two entries, which give the number of marked and unmarked intervals. The analog of equation (3.14) for impulse rewards gives

$$\Pr\{\Upsilon(t) \leq \widehat{r}\} = \sum_{n=0}^{\infty} e^{-\Lambda t} \frac{(\Lambda t)^n}{n!} \sum_{k=0}^{n+1} \widehat{\Gamma}[n, k] \Pr\{\Upsilon(t) \leq \widehat{r} | n, k\},$$

where k corresponds to the number of marked intervals. Clearly, $\Pr\{\Upsilon(t) \leq \widehat{r} | n, k\} = 1$ if we have $k \leq \widehat{r}$, while it is 0 otherwise. Furthermore, note that at most n intervals can be marked, since only n transitions can occur (the first interval is not marked). Therefore,

$$\Pr\{\Upsilon(t) \le \hat{r}\} \;=\; \sum_{n=0}^{\infty} e^{-\Lambda t}\frac{(\Lambda t)^n}{n!}\sum_{k=0}^{\lfloor\min(n,\hat{r})\rfloor}\widehat{\Gamma}[n,k]. \tag{3.15}$$

In order to evaluate (3.15), it remains to calculate $\widehat{\Gamma}[n,k]$. The entries of the matrix $\widehat{R}$ are either 0 or 1, and $\hat{r}_{ij} = 1$ if the transition from i to j corresponds to an event of interest, while $\hat{r}_{ij} = 0$ otherwise. Let $\widehat{\Gamma}_j[n,k]$ be the probability that, given n transitions, k marked events occur and the last transition is to state j. Then

$$\widehat{\Gamma}_j[n,k] \;=\; \sum_{i:\hat{r}_{ij}=1} \widehat{\Gamma}_i[n-1,k-1]p_{ij} + \sum_{i:\hat{r}_{ij}=0}\widehat{\Gamma}_i[n-1,k]p_{ij},$$

where the counter k is incremented if the transition from i to j is marked. Finally

$$\widehat{\Gamma}[n,k] \;=\; \sum_{j=1}^{M}\widehat{\Gamma}_j[n,k].$$

We now consider a more complicated measure to compute. Assume that reward rates are associated with states and that our interest is in calculating the distribution of the total reward accumulated during the observation period. In this case, $M(t) = \Pr\{Y(t) \le r\}$.

Assume that n transitions of the uniformized chain occur during a finite interval $(0,t)$. Recall that these transitions are governed by a Poisson process $\mathcal{N}$ with rate Λ, and the state of the system between transitions is given by a discrete-time Markov chain $\mathcal{Z}$. Further, $\mathcal{N}$ and $\mathcal{Z}$ are independent. The n events occur at times $\tau_1 < \tau_2 < \cdots < \tau_n$, which split the period $(0,t)$ into $n+1$ subintervals with lengths $T_1, T_2, \ldots, T_{n+1}$. Let $U_1, U_2, \ldots, U_n$ be independent and identically distributed random variables uniform on $(0,1)$, and let $U_{(1)}, U_{(2)}, \ldots, U_{(n)}$ be their order statistics. Conditioning on n events by time t, the τ_i have the same distribution as the $tU_{(i)}$. Therefore, we can identify

$$
\begin{aligned}
T_1 &= tU_{(1)} \\
T_2 &= t(U_{(2)} - U_{(1)}) \\
&\;\;\vdots \\
T_{n+1} &= t(1 - U_{(n)}).
\end{aligned}
$$

Although the T_i are dependent random variables, they are *exchangeable*. That is,

$$\Pr\{T_1 \le s_1, \ldots, T_{n+1} \le s_{n+1}\} = \Pr\{T_{l_1} \le s_1, \ldots, T_{l_{n+1}} \le s_{n+1}\}$$

for all permutations $l_1, l_2, \ldots, l_{n+1}$ of $1, 2, \ldots, n+1$ [390]. The following properties are a direct consequence of the exchangeability of the interval lengths T_i: (1) any two sequences of $k \le n$ of the T_i have the same joint distribution; (2) the sum of any $k \le n$ of the T_i has the same distribution as $tU_{(k)}$. This allows a set of k intervals with the same reward to be thought of as occurring consecutively, so that the distribution of the sum of their lengths is given by that of $tU_{(k)}$.

Let $\|\mathbf{k}\| = n+1$, and define $G_{\mathbf{k}} \subset \mathcal{S}^{n+1}$ to be the set of all vectors of states of $\mathcal{Z}$ of length $n+1$ that yield the colouring vector $\mathbf{k}$, i.e., those for which k_l of the states have reward r_l^*, $l = 1, \ldots, K+1$. Consider $\boldsymbol{\nu} \in G_{\mathbf{k}}$, where $\boldsymbol{\nu} = \langle \nu_1, \ldots, \nu_{n+1} \rangle$, and $\nu_i \in \mathcal{S}$ corresponds to the ith subinterval of $(0, t)$. Given n transitions during $(0, t)$, the colouring $\mathbf{k}$, and the "sample path" $\boldsymbol{\nu}$, then

$$Y(t)|n, \mathbf{k}, \boldsymbol{\nu} = \sum_{i=1}^{n+1} r_{\nu_i} T_i, \tag{3.16}$$

where recall that r_{ν_i} is the reward rate associated with the state ν_i and T_i is the length of the ith subinterval. From the exchangeability results of the T_i and the fact that $\mathcal{Z}$ is independent of the associated Poisson process $\mathcal{N}$, it can be shown that $Y(t)|n, \mathbf{k}, \boldsymbol{\nu}$ is independent of the particular value of $\boldsymbol{\nu}$ that yields the colouring $\mathbf{k}$ and depends only on the number of times each colour appears. Furthermore, one can assume that the first k_1 intervals have reward r_1^*, the next k_2 intervals have reward r_2^*, and so on.

We first consider the case where only two reward rates are present as in [103]. Then we have $r_1^* = 1$ and $r_2^* = 0$. In availability models, a reward rate of 1 is assigned to operational states, while a 0 reward rate is assigned to states representing a failed system. In this case $Y(t)$ is the total time during $(0, t)$ spent in the subset of states associated with reward 1, i.e., the operational time. A colouring vector $\mathbf{k}$ with $\|\mathbf{k}\| = n+1$ has two entries, but the first entry determines the second. For the measure $M(t) = \Pr\{Y(t) > r\}$, equation (3.14) can be rewritten as

$$M(t) = \sum_{n=0}^{\infty} e^{-\Lambda t} \frac{(\Lambda t)^n}{n!} \sum_{k=1}^{n+1} \Gamma[n, k] M(t, n, k), \tag{3.17}$$

where $\Gamma[n, k]$ is the probability that k of the $n + 1$ intervals are associated with reward 1. Since $r \geq 0$, the term for $k = 0$ does not appear, i.e., $M(t, n, 0) = 0$.

The conditional measure $M(t, n, k)$ is the distribution of the sum of the lengths of the k intervals associated with reward 1. The arguments about exchangeability of the intervals T_i presented above imply that $Y(t)|n, k$ has the same distribution as $tU_{(k)}$. Therefore, we have

$$M(t, n, k) = \Pr\{tU_{(k)} > r\} = \sum_{i=0}^{k-1} \binom{n}{i} \left(\frac{r}{t}\right)^i \left(1 - \frac{r}{t}\right)^{n-i}. \qquad (3.18)$$

Define $\Gamma_i[n, k]$ as the probability that k out of $n+1$ intervals are associated with reward 1 and the state visited in the last transition is i. Also define

$$\boldsymbol{\Gamma}[n, k] = \langle \Gamma_1[n, k], \ldots, \Gamma_M[n, k] \rangle.$$

Then we have

$$\Gamma[n, k] = \|\boldsymbol{\Gamma}[n, k]\| = \sum_{i=1}^{M} \Gamma_i[n, k].$$

Assume that the states in the model are organized such that the first L are associated with reward $r_1^* = 1$ and the last $M - L$ are associated with reward $r_2^* = 0$. Define

$$\begin{aligned}
\boldsymbol{\Gamma}_O[n, k] &= \langle \Gamma_1[n, k], \ldots, \Gamma_L[n, k] \rangle \\
\boldsymbol{\Gamma}_F[n, k] &= \langle \Gamma_{L+1}[n, k], \ldots, \Gamma_M[n, k] \rangle.
\end{aligned}$$

Then $\boldsymbol{\Gamma}[n, k]$ can be evaluated from the recursion (see [103] for details)

$$\begin{aligned}
\boldsymbol{\Gamma}_O[n, k] &= \boldsymbol{\Gamma}[n - 1, k - 1]\mathbf{P}_O \\
\boldsymbol{\Gamma}_F[n, k] &= \boldsymbol{\Gamma}[n - 1, k]\mathbf{P}_F.
\end{aligned} \qquad (3.19)$$

Here $\mathbf{P}_O$ and $\mathbf{P}_F$ are the submatrices of $\mathbf{P}$ consisting of its first L columns and last $M - L$ columns, respectively. In [420] $\boldsymbol{\Gamma}[n, k]$ is obtained in an explicit matrix form.

In order to calculate $\Pr\{Y(t) > r\}$, the infinite sum in (3.17) must be truncated. The error introduced by truncating to N steps can be easily bounded as in (3.4). It was noted in [103] that the sum on k in (3.17) can also be truncated, and for the class of models of highly available systems, only a few terms in the sum are necessary to obtain an accurate answer. The reasoning behind this observation is that, for availability models, most of the time is spent in states associated with $r_1^* = 1$, and so the probability that k intervals out of $n + 1$ have reward 1 should be negligible for values of k that are not near $n + 1$. Therefore, truncating by column as in [103], we can approximate (3.17) by

$$M(t) \approx \sum_{n=0}^{N} e^{-\Lambda t} \frac{(\Lambda t)^n}{n!} \sum_{k=\max(0,n-C)+1}^{n+1} \Gamma[n,k] \sum_{i=0}^{k-1} \binom{n}{i} \left(\frac{r}{t}\right)^i \left(1 - \frac{r}{t}\right)^{n-i},$$

$$(3.20)$$

where the column index $C \ll N$ for this broad class of models. This leads to computational savings in evaluating (3.17). As was the case for $\varepsilon(N)$, the error of this column truncation can also be bounded. However, the value of C was not obtained in advance.

Rubino and Sericola [397] developed a clever way of obtaining a column truncation value ahead of time such that the prespecified error bound can be achieved. We now present a slightly modified version of their approach. Suppose that a truncation is performed as discussed above for a column value C_1. From (3.20), it can be verified that the error introduced by this truncation is

$$\varepsilon^*(N, C_1) = \sum_{n=C_1+1}^{N} e^{-\Lambda t} \frac{(\Lambda t)^n}{n!} \sum_{k=1}^{n-C_1} \Gamma[n,k] \sum_{i=0}^{k-1} \binom{n}{i} \left(\frac{r}{t}\right)^i \left(1 - \frac{r}{t}\right)^{n-i}.$$

Exchanging the order of summation on k and i and making the change of variable $j = n - i$, it follows that

$$\varepsilon^*(N, C_1) \leq \sum_{n=C_1+1}^{N} \sum_{j=C_1+1}^{n} e^{-\Lambda t} \frac{(\Lambda t)^n}{n!} \binom{n}{j} \left(\frac{r}{t}\right)^{n-j} \left(1 - \frac{r}{t}\right)^{j} \sum_{k=1}^{n-C_1} \Gamma[n,k].$$

$$(3.21)$$

The last sum in (3.21) is clearly bounded by 1. Now we can write $e^{-\Lambda t} = e^{-\Lambda(r/t)t} e^{-\Lambda(1-r/t)t}$ and $(\Lambda t)^n = (\Lambda t)^j (\Lambda t)^{n-j}$, and after exchanging the order of summation on n and j in (3.21) we obtain

$$\varepsilon^*(N, C_1) \leq \sum_{j=C_1+1}^{N} e^{-\Lambda t(1-r/t)} \frac{[\Lambda t(1 - r/t)]^j}{j!} \sum_{n=j}^{N} e^{-\Lambda t(r/t)} \frac{[\Lambda t(r/t)]^{n-j}}{(n - j)!}.$$

$$(3.22)$$

Therefore, since the second sum in (3.22) is bounded by 1, we have

$$\varepsilon^*(N, C_1) \leq 1 - \sum_{j=0}^{C_1} e^{-\Lambda t(1-r/t)} \frac{[\Lambda t(1 - r/t)]^j}{j!}. \tag{3.23}$$

For models of highly available systems, one is interested in the tail of the distribution of $Y(t)$, and so r/t should be close to 1. If this is the case, then the value of C_1 is small. Note also that C_1 can be calculated in advance from either (3.22) or (3.23), but the bound in (3.23) has fewer terms to calculate than that of (3.22) when C_1 is small. The final bound can be further improved by not discarding $\sum_{k=1}^{n-C_1} \Gamma[n, k]$ from (3.21), but then the value of C_1 cannot be specified prior to calculation.

The case with more than two colours is treated next. Suppose that $\mathbf{k}$ has $L + 1$ nonzero entries, $1 \leq L + 1 \leq K + 1$, i.e., the $n + 1$ intervals are coloured with only $L + 1$ different colours. Let $\xi(1) < \cdots < \xi(L + 1)$ be the indices of these colours. Define $n_j = \sum_{i=1}^{j} k_{\xi(i)}$ for $j = 1, \ldots, L$. Then, from (3.16) and the observations following that equation, we can write

$$Y(t)/t|n, \mathbf{k} = r^*_{\xi(1)}U_{(n_1)} + \sum_{i=2}^{L} r^*_{\xi(i)}(U_{(n_i)} - U_{(n_{i-1})}) + r^*_{\xi(L+1)}(1 - U_{(n_L)}) \tag{3.24}$$

or

$$Y(t)/t|n, \mathbf{k} = \sum_{i=1}^{L} (r^*_{\xi(i)} - r^*_{\xi(i+1)})U_{(n_i)} + r^*_{\xi(L+1)}. \tag{3.25}$$

From (3.25) it is clear that determining $\Pr\{Y(t) > r|n, \mathbf{k}\}$ is equivalent to obtaining the distribution of a linear combination of uniform order statistics on $(0, 1)$. Based on results of Weisberg [474], it can be shown (see Lemma 1 of [108]) that

$$\Pr\{Y(t)/t > r|n, \mathbf{k}\} = \sum_{i:r^*_i > r} \frac{f_i^{(k_i-1)}(r^*_i, \mathbf{k})}{(k_i - 1)!}, \tag{3.26}$$

where $f_i^{(k_i-1)}$ is the $(k_i - 1)$th derivative of the function

$$f_i(x, \mathbf{k}) = \frac{(x - r)^n}{\prod_{\substack{j=1 \\ j \neq i}}^{K+1} (x - r^*_j)^{k_j}}. \tag{3.27}$$

Using (3.26) in Equation (3.14) gives

$$\Pr\{Y(t)/t > r\} = \sum_{n=0}^{\infty} e^{-\Lambda t} \frac{(\Lambda t)^n}{n!} \sum_{\mathbf{k} \in \mathcal{K}_n} \Gamma[n, \mathbf{k}] \sum_{i:r_i^* > r} \frac{f_i^{(k_i-1)}(r_i^*, \mathbf{k})}{(k_i - 1)!} \qquad (3.28)$$

where $\mathcal{K}_n = \{\mathbf{k} : ||\mathbf{k}|| = n + 1\}$.

It is possible to develop recursions to calculate $\Gamma[n, \mathbf{k}]$ and the derivatives of the functions $f_i(x, \mathbf{k})$. Note that the sum on $\mathbf{k}$ in (3.28) leads to computational requirements that are combinatorial with the number of different reward rates in the model. However, recursions which drastically reduce the computational complexity have been developed by aggregating terms in the sum on $\mathbf{k}$ [108], and we now briefly describe that approach. Specifically, let $G_g[i, n] = \{\mathbf{k} \in \mathcal{K}_n : k_i = g\}$, i.e., the set $G_g[i, n]$ contains all colouring vectors $\mathbf{k} \in \mathcal{K}_n$ that represent all paths with g intervals associated with reward r_i^*. Using recursions based on the sets $G_g[i, n]$, it can be shown that

$$\Pr\{Y(t)/t > r\} = \sum_{n=0}^{\infty} e^{-\Lambda t} \frac{(\Lambda t)^n}{n!} \sum_{i:r_i > r} ||\Upsilon[i, n, n]||, \qquad (3.29)$$

where $\Upsilon[i, n, n]$ is a vector of length M, the size of the state space. Identical expressions also hold for $\Pr\{Y(t)/t = r\}$ and $\Pr\{Y(t)/t < r\}$, where $>$ in (3.29) is replaced with $=$ or $<$ as appropriate. Simple recursions exist for $\Upsilon[i, n, u]$ in terms of the transition matrix $\mathbf{P}$, and the resulting computational complexity is linear in the number of reward rates. In fact, for many performability models, we wish to evaluate the tail of the distribution of $Y(t)$. For instance, our interest may be in the probability that the total accumulated reward is close to the maximum reward that can be achieved, and in this case the number of terms in the summation on i in (3.29) is equal to or near 1. The case for which only impulse rewards occur in the model is also considered in [108].

In [342] Nabli and Sericola obtained another polynomial algorithm in the rate reward case. After uniformizing the chain, they conditioned on the time τ_1 of the first transition as in [135], which yields an integral renewal equation. A recursion is then found after integrating the resulting equation by parts. Specifically, they considered the complementary distribution $\tilde{F}(y, t) = \Pr\{Y(t) > y\}$, which satisfies an equation similar to (3.12), and showed, given n transitions in $(0, t)$, that

$$\tilde{F}(y, t|n) = \sum_{k=0}^{n} \sum_{j=1}^{m} \binom{n}{k} y_j^k (1 - y_j)^{n-k} b^{(j)}(n, k) \mathcal{I}\{r_{j-1}t \le y < r_j t\}. \quad (3.30)$$

Here $y_j = (y - r_{j-1}t)/(r_j t - r_{j-1}t)$, $\mathcal{I}$ is an indicator random variable and the coefficients $b^{(j)}(n, k)$ are given recursively. One advantage of this method over previous algorithms is that the recursion presented involves only real numbers in the interval $[0, 1]$.

In [134] it is shown that the algorithm of [342] can be obtained using the methodology of [108] outlined above by using an efficient recursion for a linear combination of order statistics based on equation (3.27). Furthermore, a slightly simplified recursion than that of [342] is obtained. Specifically, for a given reward r_l, we have

$$\Pr\{Y(t)/t > r_l\} = \sum_{n=0}^{\infty} e^{-\Lambda t} \frac{(\Lambda t)^n}{n!} \|\Upsilon(n, 0, l - 1)\|, \quad (3.31)$$

where $\Upsilon(n, m, j)$ is computed using the recursion

$$\Upsilon_s(n, m, j) = \begin{cases} p(s, j)\Upsilon_s(n, m - 1, j) + \\ (1 - p(s, j))\sum_{s' \in S} \Upsilon_{s'}(n - 1, m - 1, j)p_{s',s}, & r_{c(s)} \ge r_j, \\ q(s, j)\Upsilon_s(n, m + 1, j) + \\ (1 - q(s, j))\sum_{s' \in S} \Upsilon_{s'}(n - 1, m, j)p_{s',s}, & r_{c(s)} < r_j. \end{cases}$$
$$(3.32)$$

Here $p(s, j) = (r_{c(s)} - r_j)/(r_{c(s)} - r_{j+1})$ and $q(s, j) = (r_{j+1} - r_{c(s)})/(r_j - r_{c(s)})$. The case for which the reward level is not in the set of the $K + 1$ distinct rewards of the model can be easily handled by including r as a fictitious reward that is not associated with a state; details are in [134].

The above recursions apply to the case for which the accumulated reward in the interval of interest is not limited by any predetermined values. However, it may be useful in certain cases to limit the amount of reward that can be accumulated in the observation interval. Assume, for example, that the reward rate for state i is $\lambda_i - C$, where λ_i is the arrival rate and C is the capacity of the channel. In this case, if the total reward $Y(t)$ is constrained to remain between 0 and a given positive value B, then $Y(t)$ represents the queue length at time t of a system with a finite buffer B. Sericola [421] adapted the algorithm of [342] and considered the case for which $Y(t)$ was bounded below by 0 but was not bounded above. The resulting solution was used to solve queueing models with an infinite buffer. In [113, 53] it was shown how to use equations (3.31) and (3.32) when both upper and lower bounds on $Y(t)$ are given. The results obtained are useful for traffic engineering.

Qureshi and Sanders [378] were the first to consider models for which both rate and impulse rewards are present. They followed the development in [104] to calculate the distribution of cumulative reward in such models, but their resulting algorithm is combinatorial in the number of reward rates. In [109] (see also [111]) the approach leading to (3.29) was extended to include models with both rate and impulse rewards, and the resulting algorithm has efficient computational requirements similar to that of [108].

Several other performability measures have also been studied using uniformization. For instance, in [99] the random variables of interest are the cumulative time in a subset of states (say, the operational states) and the number of visits to the complementary subset (say, the failed states), and their joint distribution is found. The measure of interest is the probability that the operational time exceeds a given value and the number of system failures is less than k. In [395] the distribution of the kth sojourn time in a given subset of states is obtained. The measure of interest is the probability that the total operational time between the kth and $(k+1)$th failures is greater than a given value.

3.5 Conclusions

Uniformization is a very useful technique for calculating a number of transient and steady state performability measures. One of the major advantages of uniformization is its probabilistic interpretation, which can be used to derive various results. We have presented the foundations of the method and discussed the main issues involved in the calculations. We have also discussed extensions to the basic approach and algorithms based on the technique that were developed to calculate important performability measures.

Although uniformization is one of the best methods for transient analysis, many research issues still require further investigation. Despite the recent progress that has been made, there is still room for improvements in uniformization-based methods in areas such as chains with large Λt values, nonhomogeneous chains, semi-Markov processes, and calculating probability distributions, among others. Nevertheless, the uniformization method is applicable to many problems, is easy to implement and is numerically robust.

Chapter 4

Closed-Form Solutions for Performability

Lorenzo Donatiello
Vincenzo Grassi

Markov reward models are often used for performability analysis. The aim of this chapter is to investigate the mathematical properties of these models and the methods used to calculate closed-form solutions of measures derived from them such as the distribution and moments of the cumulative reward. The investigation is mainly based on a classical functional equation obtained from these models by a renewal theoretic argument. The same equation is the basis for most of the solution methods that have appeared to date. By looking at the obtained closed-form solutions, it can be observed that the classes of reward processes they refer to have a common characteristic.

4.1 Introduction

IN recent years a great deal of attention has been devoted to the design and implementation of communication and processing systems which can continue to operate even in the presence of faults, albeit with degraded performance. Hence, it is important to develop modelling techniques which allow one to assess the ability of such systems to meet some overall service quality requirement in a given observation period, taking into account the alternation of different service quality levels.

After some preliminary work developed during the 1970s by several authors, these modelling techniques were unified by Meyer in the late 1970s under the general framework of *performability* modelling [315]. In this framework, the performance of a system S over a utilization period T is modelled as a random variable Y taking values in a set A of accomplishment levels. Each element of A represents a possible performance outcome which can be attained by S. The notion of "performance" used here is very general, since it refers to any possible aspect of the total system behaviour with respect to which the system's ability to perform is to be evaluated. Given this model, the *performability* of S is defined as:

> $Perf(B) = \Pr\{Y \in B\}$ is the probability that S performs at a level in B where B is any measurable subset of A. Performability is evaluated by solving a stochastic process $X = \{X(\tau), \tau \in I\}$, referred to as the base model, where $X(\tau)$ represents the state of the system S at time τ, and the index set I must include the utilization period T, (i.e. $T \subseteq I$). To solve X means to obtain from X values of $Perf(B)$ for some sets B of interest to the user.

From this brief description of a performability model, it should be clear that the base model X includes as special cases both queueing models commonly used in performance evaluation and Markov and semi-Markov models commonly employed in reliability/availability evaluation. In its more general form, X can represent simultaneous variations in the system's components and environment state. Hence, it is well suited for the evaluation of degradable systems where changes in structure, internal state and environment can all have an influence on the system's ability to perform.

As can be easily understood, to solve such a general and powerful model can be a formidable task. A common approach used to reduce the complexity of the solution is based on a behavioural decomposition. Indeed, in the case of gracefully degradable computer systems, two types of events can be observed: *performance events* causing performance state changes, and *dependability events* causing changes in the structure of the system.

Performance events take place far more often than dependability events. Hence, it seems reasonable to assume that the performance of the system between successive dependability state changes is in steady state most of the time. Thus, the percentage of time for which the system actually is not in steady state between successive structure state changes is assumed to be negligible. This means that the overall base model X can be decomposed into a *structural model* used to describe changes in the system structure, and a family of *performance models* (in general, one for each possible configuration that the system structure can assume) which describe the performance behaviour of the system. Such performance models can be solved for their steady state (which is usually simpler to deal with, in comparison to transient state), and then the information so obtained must be conveyed in some way to the structural model. The resulting combined model can thus be solved to obtain steady-state or transient performability measures. The model which is mostly used within this framework consists of a Markov or semi-Markov reward process. For this reason, in this chapter we focus our attention on (semi-)Markov reward processes only. However, the modeller must be warned that they are approximate models, whose validity depends on the degree of approximation introduced by the assumption that the performance measure of interest is in steady state between successive structure state changes. The closer this assumption is to reality, the more valid are the obtained results.

The chapter is organized as follows. In Section 4.2 reward models are formally defined, and two performability measures based on these models are introduced. In Section 4.3 we focus our attention on one of these two measures (the cumulative reward) and discuss the basic properties of the mathematical model which is to be solved. In Section 4.4 we analyze techniques used to get closed-form solutions of the mathematical model discussed in the previous section. In Section 4.5 we discuss an alternative way which has been exploited to solve reward models, and analyze results obtained within this framework along the lines used in Section 4.4. Finally, in Section 4.6 we summarize the results provided in the previous sections and discuss some peculiar characteristics of these results.

4.2   Reward models

We consider a system which during a given observation period $[0, t]$ can assume different configurations because of changes in its structure (caused, for instance, by faults and repairs of components). By assuming that the time spent in each system configuration is a random variable independent of past configurations, the system state at instant $\tau \in [0, t]$ can be represented by the state of a homogeneous semi-Markov process $\mathbf{X} = \{X(\tau), \tau \in [0, t]\}$, with state space $S = \{0, 1, ..., n\}$ where each $i \in S$ models a different system configuration. The dynamic behaviour of $\mathbf{X}$ is completely described by its *core matrix* $\mathbf{c}(\tau) \equiv [c_{ij}(\tau)]$, where $c_{ij}(\tau)$ is the joint probability-probability density function of a transition from state i to state j, $i, j \in S$, at instant τ [236]. Moreover, we denote by $q_i(\tau) = \sum_{j=0}^{n} c_{ij}(\tau)$ the unconditional state occupancy time density for state i, and by $Q_i(\tau) = \int_0^\tau q_i(s)ds$ the associated probability distribution.

If $\mathbf{X}$ is a Markov chain (which represents the model used most frequently) then $c_{ij}(\tau) = p_{ij}\lambda_i e^{-\lambda_i \tau}$ and $Q_i(\tau) = 1 - e^{-\lambda_i \tau}$, where p_{ij} is the probability of a transition from state i to j upon the occurrence of a transition out of state i.

Whenever the system is in state $i \in S$, a reward is accumulated at a rate $r(i)$, where r denotes a reward function $r : S \dashrightarrow R$. Each element in the set $R = \{\rho_0, \rho_1, ...\rho_m\}, m \leq n$, represents a possible performance level which can be provided by the system.

The semi-Markov process $\mathbf{X}$ together with the reward function $r(.)$ defines a *semi-Markov reward process* $[\mathbf{X}, r(.)]$, which plays the role of the base model X described above, where the semi-Markov process $\mathbf{X}$ corresponds to the structural model, while the reward function $r(.)$ provides the steady state performance associated with each structure state, obtained by solving an appropriate performance model.

By using this model, different variables can be defined which play the role of the above described performability variable Y. Two typical random variables commonly adopted are the following:

- $R_i(t)$: the instantaneous reward rate at instant t, given that the initial state was $i \in S$;

- $Y_i(t)$: the total reward accumulated during the time interval $[0, t]$, given that the initial state was $i \in S$.

From their definition, we have that these variables, in terms of the base model $[\mathbf{X}, r(.)]$, are defined as $R_i(t) = r(X_i(t))$ and $Y_i(t) = \int_0^t r(X_i(\tau))d\tau$, where $X_i(\tau)$ denotes the state of the semi-Markov process $\mathbf{X}$ at instant τ given that $X(0) = i$.

4.3   Properties of reward model solutions

When the performability base model is a semi-Markov reward process $[\mathbf{X}, r(.)]$, its solution corresponds to determining the distribution function of a performability variable defined on this model. Here we consider only the two variables defined above, $R_i(t)$ and $Y_i(t)$. In some applications, a less complete solution (for instance, some of the first moments) is sufficient. In both cases, the solution can be obtained under steady state or transient conditions, where solutions for the former are generally simpler to compute. We use the symbol "∞" instead of "t" to denote a steady state solution. Hence, we are taking into consideration the following variables: $R_i(\infty), R_i(t), Y_i(\infty)$, and $Y_i(t)$.

It can be shown [27, 81, 463] that the solution of the reward model $[\mathbf{X}, r(.)]$ for the performability variables $R_i(\infty), R_i(t)$ and $Y_i(\infty)$ requires only the evaluation of the state probabilities (transient or steady-state) of the process $\mathbf{X}$, or of a process obtained by an opportune modification of $\mathbf{X}$. This represents a classical problem of stochastic analysis, and several solution techniques have been devised for this purpose. On the other hand, the solution for $Y_i(t)$ has been revealed to be quite a difficult task, and special solution methods are needed for this purpose. Hence, the solution concerning $Y_i(t)$ represents the most interesting case, so we further limit our attention to solutions for $Y_i(t)$ only.

In the following, we will denote by $F(i, t, y) \equiv \Pr\{Y_i(t) \leq y\}$ and $M(i, t, m) \equiv E[(Y_i(t))^m]$ the probability distribution function and the mth moment of $Y_i(t)$, respectively.

A common way to attack the problem of finding the solution of a reward model $[\mathbf{X}, r(.)]$ is to write a functional equation involving the measure of interest (in this case, $F(i, t, y)$ or $M(i, t, m)$). This should at least provide a way to reason about the properties of such a solution, and, hopefully, could suggest a method of calculating the actual solution. In the case we are considering, a renewal theoretic argument can be used to write such an equation: if we consider the instant of first occurrence of a transition of the process $\mathbf{X}$ from state i, it can occur either after the instant t, or before it. If it occurs after t, the reward accumulated during $[0, t]$ is equal to $r(i)t$; if it occurs at an instant $\tau < t$, towards a state j, then the accumulated reward is $r(i)\tau$ plus the reward accumulated in the interval $[\tau, t]$, starting from state j. This leads to the following equation for $F(i, t, y)$ (see, for instance, [240]):

$$F(i, t, y) = \mathbf{1}(y - r(i)t)(1 - Q_i(t)) + \sum_{j=0}^{n} \int_0^t c_{ij}(\tau)F(j, t - \tau, y - r(i)\tau)d\tau,$$

(4.1)

where $\mathbf{1}(.)$ is the unit step function defined as:

$$\mathbf{1}(x) = \begin{cases} 1, & \text{if } x \geq 0, \\ 0, & \text{otherwise.} \end{cases}$$

The first term on the right-hand side of equation (4.1) refers to the occurrence of the first transition at an instant greater than t, while the second term refers to the occurrence of this event at any instant $\tau \in [0, t]$. It is to be noted that a first example of the functional equation concerning reward models, based on the above outlined renewal argument, can be found in [236], referring to the first moment $E[Y_i(t)]$.

If $\mathbf{X}$ is a Markov chain, then equation (4.1) takes the following form:

$$F(i, t, y) = \mathbf{1}(y - r(i)t)e^{-\lambda_i t} + \sum_{j=0}^{n} p_{ij} \int_0^t \lambda_i e^{-\lambda_i \tau} F(j, t - \tau, y - r(i)\tau)d\tau. \quad (4.2)$$

From (4.1), we can derive (see the proof in the Appendix, Section 4.7.1) an equation for the mth moment $M(i, t, m)$:

$$\begin{aligned} M(i, t, 0) &= 1, \\ M(i, t, m) &= (r(i)t)^m(1 - Q_i(t)) \\ &+ \sum_{j=0}^{n} \int_0^t c_{ij}(\tau) \sum_{k=0}^{m} \binom{m}{k} (r(i)\tau)^{m-k} M(j, t - \tau, k)d\tau. \quad (4.3) \end{aligned}$$

Actual solution methods of equations (4.1), (4.2) and (4.3) will be discussed in the next section. However, we analyze equation (4.1) to get some insight into the mathematical properties of the solution we are seeking.

Let $\mathbb{R}$ and $\mathbb{R}^+$ be the sets of real and non-negative real numbers, respectively, and let $\mathcal{B} = \{g \mid g : S \times \mathbb{R}^+ \times \mathbb{R}^+ \to \mathbb{R}\}$ be the set of functions $g(i, t, y)$ such that:

- for every $i \in S$ and for every fixed $y \in \mathbb{R}^+$, the function $t \to g(i, t, y)$ is bounded over finite intervals $[0, T]$;

- for every $i \in S$ and for every fixed $t \in \mathbb{R}^+$, the function $y \to g(i, t, y)$ is bounded;

- for every fixed $y \in \mathbb{R}^+$ and $t \in \mathbb{R}^+$, the function $i \to g(i, t, y)$ is bounded over finite intervals.

Note that the cumulative reward distribution $F(i, t, y) \in \mathcal{B}$ and is non-negative. A function $g \in \mathcal{B}$ is said to satisfy a *Markov reward renewal equation* if for all $i \in S$ and $t, y \in \mathbb{R}^+$:

$$g(i, t, y) = h(i, t, y) + \sum_{j \in S} \int_0^t c_{ij}(\tau) g(j, t - \tau, y - r(i)\tau) d\tau, \qquad (4.4)$$

for some non-negative function $h \in \mathcal{B}$. Equations (4.1) and (4.2) are in the form of (4.4), since $\mathbf{1}(y - r(i)t)(1 - Q_i(t)) \in \mathcal{B}$ and is non-negative.

Equation (4.4) is a generalization of the *Markov renewal equation* [87] and has a unique solution when, as in the case we are considering here, the state space S is finite. In [87] some sufficient conditions for the uniqueness of the solution, when the state space S is infinite, are also given.

To conclude this section, we present an interesting mathematical property which characterizes equations (4.1) and (4.3). Based on such a property, we outline a numerical procedure to obtain the moments and distribution of $Y_i(t)$. The procedure allows us to obtain also an upper bound for the introduced approximation error. For the sake of conciseness, let us denote by $\phi : \mathcal{B} \to \mathcal{B}$ the functional defined by the right-hand side of equation (4.1), i.e.:

$$\phi(g(i, t, y)) \equiv \mathbf{1}(y - r(i)t)(1 - Q_i(t))$$
$$+ \sum_{j \in S} \int_0^t c_{ij}(\tau) g(j, t - \tau, y - r(i)\tau) d\tau, \quad g \in \mathcal{B}. \quad (4.5)$$

Let $d : \mathcal{B} \times \mathcal{B} \to \mathbb{R}$ be a "distance" function defined as:

$$d(v, w) \equiv \sup_{\substack{t \in [0, T] \\ y \in \mathbb{R}, i \in S}} \{|v(i, t, y) - w(j, t, y)|\}, \quad v \in \mathcal{B}, w \in \mathcal{B}.$$

It can be shown that $(\mathcal{B}, d)$ is a *complete metric space* [117]. Given a generic function $\varphi : Z \to Z$ defined on a complete metric space (Z, d), φ is a *contraction mapping* on Z with *modulus* C if there exists a constant C, $0 \le C < 1$, such that:

$$d(\varphi(z'), \varphi(z'')) \le C d(z', z'') \quad \forall z' \in Z, z'' \in Z$$

Banach's fixed point theorem, which holds for contraction mappings, states the following [117].

1. There exists a unique solution $\zeta \in Z$ for the fixed point equation $z = \varphi(z)$.

2. The sequence $\{z_k\}_{k \in \mathcal{N}}$ defined as:
$$\begin{cases} z_0 = z^* \text{ where } z^* \text{ may be any element of } Z \\ z_k = \varphi(z_{k-1}) \end{cases}$$
converges to ζ.

3. If the sequence is stopped at a certain term z_{k^*}, then:
$$d(\zeta, z_{k^*}) \le \frac{C^{k^*}}{1 - C} d(z_0, z_1).$$

The above inequality gives an upper bound for the approximation error incurred if we approximate ζ by means of z_{k^*}.

In the following theorem we prove a sufficient condition such that, whenever it holds, the functional ϕ defined by (4.5) is a contraction mapping on $(\mathcal{B}, d)$.

Theorem 4.1 *Let $\phi : \mathcal{B} \to \mathcal{B}$ be the functional defined by (4.5). If $Q_i(t) < 1$ for all $i \in S$ and for all $t \le T$, then ϕ is a contraction mapping on $(\mathcal{B}, d)$ with modulus $C = max_{i \in S}\{Q_i(T)\}$.*

Proof. See Section 4.7.2 in the Appendix.   □

Theorem 4.1 holds under very general conditions. Indeed, it holds certainly for many common residence time distributions (e.g., Gaussian, Weibull, etc.). If the stochastic process $\mathbf{X}$ is a Markov chain, the following corollary can be proved.

Corollary 4.1 *If* X *is a Markov chain, then* ϕ *is always a contraction mapping on* $(\mathcal{B}, d)$.

Proof. If X is a Markov chain, then $c_{ij}(\tau) = p_{ij}\lambda_i e^{-\lambda_i \tau}$, where p_{ij} is the transition probability from i to j and $\lambda_i e^{-\lambda_i \tau}$ is the state occupancy time exponential density. Hence, $Q_i(t) = 1 - e^{-\lambda_i t} < 1$, $\forall t \leq T$, and the corollary is proved. $\qquad\square$

In principle, Theorem 4.1 suggests a numerical procedure for approximating, with known bound, the exact solution of equation (4.1):

- The sequence $\{F^{(k)}\}_{k\in\mathcal{N}}$ defined as:

$$\begin{cases} F^{(0)}(i,t,y) = 0 \ \text{(where 0 is the null function)} \\ F^{(k)}(i,t,y) = \phi(F^{(k-1)}(i,t,y)) \end{cases}$$

  converges to $F(i,t,y)$.

- The distance between $F^{(k)}$ and F is upper bounded by:

$$d(F^{(k)}, F) \leq \frac{(\max_i\{Q_i(T)\})^k}{1 - \max_i\{Q_i(T)\}} d(F^{(0)}, F^{(1)}) \leq \frac{(\max_i\{Q_i(T)\})^k}{1 - \max_i\{Q_i(T)\}}.$$

A similar result can be proved for equation (4.3). It must be noted that the above-defined upper bound holds only if the integral defined in the right-hand side of (4.1) is exactly evaluated at each iteration step. Otherwise (as in the case of numerical integration techniques) the effect of the evaluation error must be taken into account.

However, as far as we know, this approach has not been investigated, and the existing numerical solutions are based on different approaches [103, 104, 187, 370, 384, 388, 440].

4.4 Definition and genesis of closed-form solutions

In the previous section we have proven that the solution of equation (4.1) exists and is unique when the state space S is finite. Moreover, we have also provided an equation for the mth moment of $Y_i(t)$. In this section we discuss methods to obtain explicit solutions for $F(i,t,y)$ and $M(i,t,m)$.

In general, the solution of a mathematical model can be given either
as a *closed-form solution* or as a *numerical solution*. By a closed-form solution we mean a solution expressed as an explicit function of the model
parameters, which can be "exactly" evaluated. This definition excludes,
for instance, solutions expressed as integrals of functions, where the integration cannot be performed analytically, or as infinite summations whose
limit is not known.[1]

On the other hand, a numerical solution basically consists of a procedure for deriving, given a particular set of values of the model parameters,
the corresponding numerical value of the solution.

In this chapter, we focus our attention on closed-form solutions of reward models. The approaches adopted to get such solutions can be grouped
into two classes, which we call analytical-probabilistic approaches and direct probabilistic approaches. These two approaches will be discussed in
the following two subsections, after which we discuss the involved computational complexity.

4.4.1   Analytical-probabilistic approaches

Equation (4.1), which has been used in Section 4.3 to obtain insight into
the mathematical properties of the reward models which are to be solved,
has been actually used as the starting point of most of the solution methods
presented in the literature.

An approach which has proven successful is based on the transformation
of the equation to a transformed domain, where it assumes a more manageable form from which a closed-form solution can be calculated (in the
transform domain) with relative ease, and then inverted to get the actual
solution in the starting domain. For existing solutions the adopted domain
is the Laplace transform domain, and this approach has been mostly used
when $\mathbf{X}$ is a Markov chain, i.e. when $c_{ij}(\tau) = p_{ij}\lambda_i e^{-\lambda_i \tau}$. The approach
works as follows.

Let $L(i, u, v) = \int_{t=0}^{\infty} \int_{y=0}^{\infty} f(i, t, y) e^{-vy} e^{-ut} \, dy \, dt$ be the double Laplace-
Stieltjes and Laplace transform of $F(i, t, y)$ in the y and t variable, respec-

---

[1] Actually, little agreement can be found in the literature about what is and what is
not a closed-form solution. This problem could give rise to a somewhat "phylosophical"
discussion which is outside the scope of this chapter. Hence, we have chosen the above-
mentioned "pragmatic" approach.

tively (where $f(i, t, y)$ denotes the probability density function associated with $F(i, t, y)$). We obtain from (4.1):

$$L(i, u, v) = \hat{Q}_{ij}(u + vr(i)) + \sum_{j=0}^{n} \hat{c}_{ij}(u + vr(i))L(j, u, v) \quad (4.6)$$

where $\hat{Q}(.)$ and $\hat{c}_{ij}(.)$ are the Laplace transform of $(1 - Q_i(t))$ and $c_{ij}(t)$, respectively. If $\mathbf{X}$ is a Markov chain, then equation (4.6) takes the form:

$$L(i, u, v) = \frac{1}{\lambda_i + u + r(i)v} + \frac{\lambda_i}{\lambda_i + u + r(i)v} \sum_{j=0}^{n} p_{ij} L(j, u, v) \quad (4.7)$$

By comparing equation (4.1) and equation (4.6) or (4.7), we see that the original integral equation with $F(i, t, y)$ as unknown has been transformed into a linear (and, hence, more manageable) equation with $L(i, u, v)$ as unknown.

However, even if the solution of equation (4.6) or (4.7) is relatively easy, the inversion back to the original domain is, in general, a difficult task. In the limit of our knowledge, the most general closed-form solution obtained by solving equation (4.7) and then inverting the obtained result is limited to Markov reward processes where [136, 188]:

- $p_{ij} \neq 0$, only if $i \geq j$;

- $r(i) \geq 0$.

Hence, the first condition limits the range of application of this solution to the class of models characterized by an *acyclic* structure of the transition graph. In dependability modelling, such models can be used to represent *nonrepairable* systems (a system is nonrepairable if, once it departs from a given structure state, it cannot reenter that state). The second condition is not a real limitation, since, even in the case of negative reward rates, it suffices to add an appropriate constant to make them all positive. This only causes a "shift" in the distribution to be calculated.

The actual solution presented in [188] is obtained under two further conditions, which were introduced to simplify the mathematical manipulation; their removal would simply imply a more involved closed-form expression:

- $\lambda_i \neq \lambda_j$ or $r(i) \neq r(j)$, $i \neq j$, $i, j \in S$;

- $\dfrac{\lambda_i - \lambda_j}{\lambda_k - \lambda_j} \neq \dfrac{r(i) - r(j)}{r(k) - r(j)}$, $i \neq k \neq j$, $i, k, j \in S$, $i, k \notin R_j, \notin L_j$,
  where $L_j = \{i \in S | \lambda_i = \lambda_j\}$, and $R_j = \{i \in S | r(i) = r(j)\}$.

The solution then is as follows:

$$
\begin{aligned}
F(n, t, y) \;=\;& \sum_{k=0}^{n} \sum_{\substack{i=0 \\ i \notin L_k, i \notin R_k}}^{n} \frac{H_k^{(n)}(i)}{\lambda_i - \lambda_k} e^{-\lambda_k t} \Big(1 - e^{-\frac{\lambda_i - \lambda_k}{r(i) - r(k)}(y - r(k)t\mathbf{1}(y - r(k)t))}\Big) \\
&+ \sum_{k=0}^{n} \sum_{i=1}^{M_k^{(n)}} \frac{Z_k^{(n)}(i)}{i!} e^{-\lambda_k t} (y - r(k)t)^i \mathbf{1}(y - r(k)t) \\
&+ \sum_{k=0}^{n} W_k^{(n)} e^{-\lambda_k t} \mathbf{1}(y - r(k)t),
\end{aligned}
\tag{4.8}
$$

where $M_0^{(0)} \equiv 0$,

$$
M_k^{(n)} \equiv \begin{cases} M_k^{(n-1)}, & \text{if } \lambda_n \neq \lambda_k, \\ M_k^{(n-1)} + 1, & \text{otherwise}, \end{cases}
$$

and $M_n^{(n)} \equiv \max\limits_{0 \leq k \leq n-1} \{M_k^{(n)}\}$, and the coefficients $H_k^{(n)}(i), Z_k^{(n)}(i), W_k^{(n)}$ are independent of t and y. Recursive formulas are provided to calculate these coefficients [188].

In [86] and [136] other closed-form solutions are presented, which have been obtained by using the same transform approach. They can be applied to somewhat more restrictive classes of Markov reward models, and hence can be viewed as special cases of the solution reported above.

Within the same class of analytical approaches, a different approach has been employed in [186] to get a closed-form solution of equation (4.2). The approach exploits the closure property of exponential functions with respect to convolution, so working directly in the starting domain of $F(i, t, y)$, without passing through some transform domain. The provided solution can be applied to Markov reward models where:

- $p_{ij} \neq 0$ only if $i \geq j$;

- $r(i) \geq 0, r(i) \geq r(j)$ if $i \geq j$;

- $\lambda_i \neq \lambda_j$ and $r(i) \neq r(j)$, $i \neq j$, $i, j \in S$;

- $\dfrac{\lambda_i - \lambda_j}{\lambda_k - \lambda_j} \neq \dfrac{r(i) - r(j)}{r(k) - r(j)}, \quad i \neq k \neq j, \; i, k, j \in S.$

Hence, also this solution concerns acyclic models. With respect to the solutions in [136, 188], there is the further limitation that the reward rates must be monotone in the state labelling.

Let us now consider equation (4.3). It can be solved (and indeed it has been) by using the same transform approach adopted for equation (4.1). Actually, we know only one closed-form solution obtained in this way for this equation [240]. The proposed solution is restricted to Markov reward models whose generator matrix has distinct eigenvalues. This last condition was introduced only for the sake of simplicity, and could be removed at the expense of a more involved expression. The calculated solution is:

$$\mathbf{M}(t, k) \;=\; \sum_{i=0}^{n} \sum_{j=0}^{k} \mathbf{v}_k(i, j) e^{\mu_i t} t^j, \tag{4.9}$$

where $\mathbf{M}(t, k)$ is the vector with ith entry equal to $M(i, t, k)$, μ_i is the ith eigenvalue of the process generator matrix, and $\mathbf{v}_k(i, j)$ is a vector of coefficients independent of t which can be calculated by means of recursive formulas [240]. It is to be noted that this solution requires the eigenvalues of the generator matrix to be computed. This is trivial for acyclic Markov chains (the eigenvalues appear on the diagonal), but requires specialized algorithms in the general case.

4.4.2 Direct probabilistic approaches

The other way to obtain a closed-form solution of reward models is based on the use of probabilistic techniques only. Typically, the adopted technique is the theorem of total probability [462] which, given an event A and a family of exhaustive and mutually exclusive events B_j, states that:

$$\Pr\{A\} = \sum_j \Pr\{A|B_j\}\Pr\{B_j\}.$$

By successive application of this theorem, the original problem of calculating the performability distribution (or its moments) is transformed into the problem of calculating the probabilities for a set of conditional events, and for the related conditioning events. By a careful choice of the conditioning events, this transformation actually means a reduction of the original complex problem to a set of smaller and simpler problems, each of which can be solved quite easily. This is the approach followed in [151, 316]. It is restricted to reward models where:

- $p_{ij} \neq 0$ only if $i \geq j$;

- $r(i) \geq 0, r(i) \geq r(j)$ if $i \geq j$.

Hence, even with direct probabilistic approaches closed-form solutions appear restricted to the class of acyclic processes. In this restricted class of models the set U of the possible *state trajectories* visited by $\mathbf{X}$ in $[0, t]$ starting from state $i \in S$ is finite. Hence, we have:

$$F(i, t, y) = \sum_{u \in U} F(i, t, y|u) \Pr\{u\}.$$

For each possible $u \in U$, the corresponding $\Pr\{u\}$ can be easily evaluated from the discrete time Markov chain underlying the continuous time process $\mathbf{X}$. On the other hand, the conditional cumulative reward distribution $F(i, t, y|u)$ can be calculated by further applying the theorem of total probability:

$$F(i, t, y|u) \quad = \quad \int \ldots \int_{C_y} f(\mathbf{v}|u) dv_1 \ldots dv_n, \tag{4.10}$$

where $f(\mathbf{v}|u)$ is the conditional joint density function of the random vector $\mathbf{v} \in V$, the set V is defined as $V = \{\mathbf{v}|v_i \equiv \text{time spent in } [0, \infty] \text{ in state } i \text{ by the process } \mathbf{X}\}$, and C_y is the integration region used to take into account the fact that the actual observation period is $[0, t]$. To obtain a closed-form solution, the multiple integral in (4.10) must be explicitly calculated, which is generally a difficult task. If $\mathbf{X}$ is a Markov chain, then only exponential functions are involved, and the integration can be performed (possibly, with the aid of a symbolic manipulation tool). Actually, an explicit closed-form expression has been provided in [316] only for the case of a three-state Markov chain.

Another closed-form solution which has been presented refers to a very limited class of semi-Markov reward models, with state space $S = \{0, 1\}$, where $r(0) = 0$, $r(1) = 1$, and the residence time in state 0 is constant equal to c, while the residence time in state 1 is exponential with parameter λ. However, it is to be noted that in this special case the solution trivially extends to any possible value of the reward rates. By using simple probabilistic arguments, the model solution can be calculated as (see [396]):

$$F(i, t, y) = 1 - e^{-\lambda y} \sum_{i=0}^{\lfloor (t-y)/c \rfloor} \frac{(\lambda y)^i}{i!}.$$

Finally, in the limit of our knowledge, direct probabilistic methods for the solution of equation (4.3) have not been presented in the literature.

4.4.3 Computational complexity

The closed-form solutions calculated by the two approaches outlined above can be compared in terms of their computational complexity.

With regard to the solution (4.10) presented in [151, 316], it can be easily realized that the set U of the state trajectories has a cardinality which is exponential in the cardinality of the state space S. Hence, the computational complexity of this solution is at least $O(a^n)$, for some constant $a > 1$, where $n = |S|$. This means that this solution can be actually used for small size models only.

On the other hand, it has been shown in [188] that the computational complexity of (4.8) is $O(\eta n^2)$, where η is the number of nonzero entries in the transition matrix of $\mathbf{X}$. Hence, in the worst (and very rare) case of a fully connected transition graph for $\mathbf{X}$, the computational complexity is $O(n^4)$. Because of its polynomial complexity, solution (4.8) is suited also for average size models. The solution in [186] has the same complexity.

The computational complexity of calculating the first k moments by means of (4.9) has been calculated as $O(k^2 n^4)$, regardless of the possible sparsity of the transition matrix. Hence, it appears greater than the complexity needed to calculate the cumulative reward distribution, with respect to the state space cardinality. This could suggest that, possibly, there is room for some improvement in the computational complexity of a closed-form solution of equation (4.3).

4.5   Alternative solution of reward models

In the previous two sections we have discussed the solution of mathematical models for performability analysis. We have focused our attention on
reward models, and have founded our analysis on equation (4.1), obtained
by using a renewal argument. As we have seen, this equation allows us
both to reason about the basic properties of the model to be solved, and
to actually calculate the solution, which, in a limited number of cases, can
be expressed in closed form.

A different approach, which does not stem from renewal arguments, has
been exploited in some papers to calculate the solution of reward models
[103, 104, 135]. This approach is intrinsically numerical since it allows only
an approximation of the required solution, while a complete closed-form
solution appears outside its scope. However, it lends itself, to some extent,
to considerations similar to those discussed in the previous sections. For
this reason, we discuss it within this context.

Let us consider the cumulative reward distribution $F(i, t, y)$ relative to
a semi-Markov reward process $[\mathbf{X}, r(.)]$ (a similar argument can be used
for the mth moment $M(i, t, m)$). We can write:

$$F(i, t, y) = \sum_{k=0}^{\infty} F(i, t, y|k)\mathrm{Pr}\{k \text{ transitions of } \mathbf{X} \text{ in } [0, t]|X(0) = i\},$$

(4.11)

where $F(i, t, y|k)$ denotes the cumulative reward distribution conditioned
on the occurrence of k transitions of $\mathbf{X}$ in $[0, t]$.

As we can see, the approach suggested by equation (4.11) to the evaluation of $F(i, t, y)$ is intrinsically numerical since, even if we are able to
calculate exactly both $F(i, t, y|k)$ and $\mathrm{Pr}\{k\}$, equation (4.11) tells us only
that the solution is the limit to which the infinite series in its right-hand
side converges. The only thing we can actually do is to try to approximate
this limit by truncating the infinite series. An interesting characteristic
of this approach is that, if $\mathrm{Pr}\{k\}$ is known, it allows an easy calculation
of an upper bound for the truncation error. If we sum only the first $\bar{k}$
terms of the infinite series, then by simple probabilistic considerations the
truncation error $\varepsilon(\bar{k})$ can be upper bounded as:

$$\varepsilon(\bar{k}) \leq 1 - \sum_{k=0}^{\bar{k}}\mathrm{Pr}\{k\}.$$

However, if we consider general reward models, even the calculation of $F(i, t, y|k)$ and $\Pr\{k\}$ is not trivial at all. A great simplification is obtained for a special class of Markov reward processes, whose exponential transition rates satisfy the following property that $\lambda_i = \gamma, \forall i \in S$, where γ is a real constant greater than zero. In this case, $\Pr\{k\}$ is trivially equal to the Poisson distribution:

$$\Pr\{k\} = e^{-\gamma t}\frac{(\gamma t)^k}{k!}.$$

To complete the solution for this class of Markov reward processes, it remains to evaluate $F(i, t, y|k)$. The only known solutions for this distribution are in closed form and, quite interestingly, they can be placed within the same framework introduced in Section 4.4 for closed-form solutions of $F(i, t, y)$. Hence, in this case too we can distinguish between solutions obtained by a direct probabilistic approach, and solutions obtained by an analytical-probabilistic approach.

The solution presented in [103, 104] belongs to the direct probabilistic class. Hence, by the theorem of total probability, $F(i, t, y|k)$ is expressed as:

$$F(i, t, y|k) = \sum_{\mathbf{h}:\ |\mathbf{h}| = k+1} \Gamma(k, \mathbf{h}) F(i, t, y|k, \mathbf{h}), \qquad (4.12)$$

where $\mathbf{h} = [h_1, \ldots, h_m]$, h_i is the number of visits to states with reward rate equal to ρ_i during $[0, t]$, $|\mathbf{h}| = h_1 + \ldots + h_m$, $\Gamma(k, \mathbf{h}) = \Pr\{\mathbf{h}|k \text{ transitions}\}$, and $F(i, t, y|k, \mathbf{h})$ is the distribution $F(i, t, y|k)$ conditioned on the occurrence of $\mathbf{h} = [h_1, \ldots, h_m]$ visits to states with reward rates $\rho_1, \ldots, \rho_m$, respectively. By using again probabilistic arguments (based on the exchangeability property [390]), recursive formulas are provided to evaluate $F(i, t, y|k, \mathbf{h})$. A recursion is also provided in [103, 104] for the evaluation of $\Gamma(k, \mathbf{h})$.

A different approach to the evaluation of a closed-form solution for $F(i, t, y|k)$ is presented in [135], which can be placed within the class of probabilistic-analytical approaches. The core of this alternative approach is based on the derivation of an integral-recurrence equation between $F(i, t, y|k)$ and $F(j, t, y|k - 1)$, $i, j \in S$. This equation is obtained by using a renewal argument based on the property that k Poisson arrivals in an interval $[0, t]$ are uniformly distributed [87]. Hence, the first of these arrivals occurs at an instant $\tau \in [0, t]$ which is distributed as the minimum of k uniform variables; on the other hand, in the remaining $[\tau, t]$ interval the stochastic process $\mathbf{X}$ behaves as in $[0, t]$, but conditioned on the occurrence of $k - 1$ rather than k transitions. The resulting equation is (see Section 4.7.3 in the Appendix for a formal proof):

$$F(i, t, y|k) = \sum_{j=0}^{n} p_{ij} \int_0^t \frac{k(t - \tau)^{k-1}}{t^k} F(j, t - \tau, y - r(i)\tau|k - 1)d\tau. \qquad (4.13)$$

The above equation is solved in [135] by using a transform approach. The obtained closed-form expression for $F(i, t, y|k)$ is:

$$\begin{aligned} F(i, t, y|k) \quad &= \quad \alpha_i^{(k)} \mathbf{1}(y - r(i)t) &\qquad (4.14) \\ &+ \quad \sum_{h=1}^{k} \sum_{w=0}^{m} \binom{k}{h-1} \beta_i^{(k)}(w, k)(\frac{y - \rho_w t}{t})^{k-h+1} \mathbf{1}(y - \rho_w t), \end{aligned}$$

where coefficients $\alpha_i^{(k)}$ and $\beta_i^{(k)}(w, h)$ are independent of t and y and can be calculated by means of recursive expressions [135].

Another probabilistic-analytical approach for the evaluation of $F(i, t, y)$ is presented in [341]. This approach considers the joint probability distribution:

$$F(i, t, y, k) \equiv \Pr\{Y_i(t) = y, k \text{ transitions in } [0, t]\},$$

rather than the conditional probability distribution $F(i, t, y|k)$ as in the previous case. Hence, it is:

$$F(i, t, y) = \sum_{k=0}^{\infty} F(i, t, y, k).$$

Then, a forward renewal equation is obtained for $F(i, t, y, k)$ and, from this equation, a closed-form expression for $F(i, t, y, k)$ is derived.

The three solutions presented above can be compared in terms of their computational complexity. The solution proposed in [103, 104] requires the evaluation of a number of terms that grows exponentially with the number m of different reward rates. This makes this solution applicable only to reward models with a limited number of reward rates (a typical case could be the models used in availability/reliability analysis, where only two different values are used). However, the solution method proposed in [103, 104] is re-elaborated in [108], leading to a complexity $O(\eta m k/2)$ for the evaluation of $F(i,t,y|k)$, where η is the number of nonzero entries of the transition matrix of $\mathbf{X}$. It should be noted that the proposed recursive formulas depend on the value of y; hence, they must be re-evaluated for each value of y one is interested in.

On the other hand, in [135] it is proven that the computational complexity for calculating $F(i,t,y|k)$ by means of (4.15), once $F(i,t,y|k-1)$ is known, is $O(\eta mk)$. Note that the evaluation procedure is independent of y, so that no significant additional computation is needed for different values of y.

In [341], it is proven that the computational complexity for calculating $F(i,t,y,k)$ is $O(\eta mk)$. With respect to [135], a more refined algorithm is developed. Such an algorithm reduces, in some cases, the number of terms to be considered to guarantee a given bound to the error (see Table 4.3 below). Moreover, the algorithm presented in [341] appears more stable, from a numerical viewpoint, with respect to the algorithms proposed in [104, 135] since it deals only with positive numbers bounded by 1.

The performance of all these algorithms degrades when the value of γ is large with respect to t, since a large number of terms is required to guarantee a given error tolerance. To overcome this problem some modifications have been proposed [59, 332]. However, such methods do not provide a closed form, so we do not consider them in this chapter.

Finally, it is to be noted that the class of Markov processes to which the approach presented in this section can be applied is only apparently limited. Actually, it has been proven that, given any homogeneous Markov chain $\mathbf{X}$, there exists always another Markov process $\mathbf{Y}$ such that $\lambda_i = \gamma, \forall i \in S$, whose probabilistic behaviour is identical to that of $\mathbf{X}$. The process of obtaining $\mathbf{Y}$ from $\mathbf{X}$ is known as "uniformization" or "randomization" [87, 198].

4.6 Conclusions

In this chapter we have considered reward processes as models for performability analysis. In particular, we have considered as measure of interest the reward accumulated over a finite time interval. The model solution with respect to such a measure consists of evaluating of its probability distribution or moments. A functional equation, obtained by a renewal argument, has been used to investigate the mathematical properties of the model, namely existence and uniqueness of the solution.

Then, we focused our attention on closed-form solutions (in the sense explained at the beginning of Section 4.4). It can be noted that, to date, such solutions are limited to a restricted class of reward models, with respect to the class for which numerical solutions have been proposed. Since it is generally harder to get a closed-form solution of a mathematical model, this should not come as a surprise.

In the case of closed-form solutions concerning the probability distribution of the cumulative reward, it is noteworthy that the models they refer to are characterized by the common property of having a finite maximum number of state transitions in the interval $[0, t]$. For the solutions in [151, 186, 188, 316] this maximum number is $n = |S|$; for the solution presented in [396] n is equal to $\lfloor t/c \rfloor$; while for the solutions in [104, 108, 135, 341], it is the problem formulation itself which limits the attention to a maximum of k transitions. It could be worth investigating whether this is a sort of "necessary" condition for having a closed-form solution.

Different closed-form solutions referring to the same class of models can be compared in terms of their computational complexity. For the sake of such comparison, we have summarized the presented solution approaches in Tables 4.1, 4.2 and 4.3, which refer to the methods presented in Sections 4.4.1, 4.4.2 and 4.5, respectively. For each method, we indicate the possible constraints on the Markov reward model, and the minimum and maximum computational cost. The maximum complexity corresponds to the case of a Markov process with "dense" transition matrix, i.e., with almost fully connected transition graph. The minimun complexity corresponds to the case where from each state transitions are possible towards at most D different states, for a "small" constant D independent of the state. We have not reported the complexity whenever it cannot be considered meaningful.

4.7 Appendices

4.7.1 Proof of equation (4.3)

Let $f(i, t, y)$ denote the density function associated with the cumulative reward distribution $F(i, t, y)$. From equation (4.1) an equation for $f(i, t, y)$

Table 4.1: Analytical-probabilistic approaches

Citation	Constraints	Complexity	
		Min.	Max.
[186]	$i < j \rightarrow p_{ij} = 0$ $\lambda_i \neq \lambda_j$ or $r(i) \neq r(j)$ $\dfrac{\lambda_i - \lambda_j}{\lambda_k - \lambda_j} \neq \dfrac{r(i) - r(j)}{r(k) - r(j)}$	$O(n^2)$	$O(n^4)$
[135]	$i < j \rightarrow (p_{ij} = 0, r(i) \leq r(j))$ $\lambda_i \neq \lambda_j$ and $r(i) \neq r(j)$ $\dfrac{\lambda_i - \lambda_j}{\lambda_k - \lambda_j} \neq \dfrac{r(i) - r(j)}{r(k) - r(j)}$	$O(n^2)$	$O(n^4)$
[81]	$i < j \rightarrow p_{ij} = 0$ $\lambda_i \neq \lambda_j$ and $r(i) \neq r(j)$ $\dfrac{\lambda_i - \lambda_j}{\lambda_k - \lambda_j} \neq \dfrac{r(i) - r(j)}{r(k) - r(j)}$	$O(n^2)$	$O(n^4)$
[108]	$(j \neq i - 1$ or $j \neq n) \rightarrow p_{ij} = 0$ $\lambda_i \neq \lambda_j$ and $r(i) \neq r(j)$ $\dfrac{\lambda_i - \lambda_j}{\lambda_k - \lambda_j} \neq \dfrac{r(i) - r(j)}{r(k) - r(j)}$	$O(n^2)$	$O(n^3)$

can be obtained by differentiation, yielding:

$$f(i, t, y) = u_0(y - r(i)t)(1 - Q_i(t)) + \sum_{j=0}^{n} \int_0^t c_{ij}(\tau) f(j, t - \tau, y - r(i)\tau) d\tau,$$

$$(4.15)$$

where $u_0(.)$ is the unit impulse function. The mth moment is defined as:

$$M(i, t, m) = \int_0^\infty y^m dF(i, t, y).$$

Then, when $m = 0$ we have:

$$M(i, t, 0) = \int_0^\infty f(i, t, y) dy = 1.$$

Table 4.2: Direct probabilistic approaches

Citation	Constraints	Complexity	
		Min.	Max.
[136, 240]	$i < j \rightarrow (p_{ij} = 0, r(i) \leq r(j))$	–	$O(a^n), a > 1$
[384]	$|S| = 2$	–	

Let us consider the case $m > 0$. By substituting the right-hand side of equation (4.15) in the definition of $M(i, t, m)$ we obtain:

$$
\begin{aligned}
M(i,t,m) &= \int_0^\infty y^m u_0(y - r(i)t)(1 - Q_i(t))dy \\
&+ \int_0^\infty y^m \sum_{j=0}^n \int_0^t c_{ij}(\tau)f(j, t - \tau, y - r(i)\tau)d\tau dy \\
&= (r(i)t)^m(1 - Q_i(t)) \\
&+ \sum_{j=0}^n \int_0^t c_{ij}(\tau) \int_0^\infty y^m f(j, t - \tau, y - r(i)\tau)dy d\tau \\
&\quad \text{(with a change of variable } z = y - r(i)\tau) \\
&= (r(i)t)^m(1 - Q_i(t)) \\
&+ \sum_{j=0}^n \int_0^t c_{ij}(\tau) \int_{-r(i)\tau}^\infty (z + r(i)\tau)^m f(j, t - \tau, z)dz d\tau \\
&= (r(i)t)^m(1 - Q_i(t)) \\
&+ \sum_{j=1}^n \int_0^t c_{ij}(\tau) \int_0^\infty (z + r(i)\tau)^m f(j, t - \tau, z)dz d\tau \\
&\quad \text{(since } f(i, t, z) = 0 \text{ for } z < 0) \\
&= (r(i)t)^m(1 - Q_i(t)) \\
&+ \sum_{j=0}^n \int_0^t c_{ij}(\tau) \int_0^\infty \sum_{k=0}^m \binom{m}{k}(r(i)\tau)^{m-k}z^k f(j, t - \tau, z)dz d\tau \\
&= (r(i)t)^m(1 - Q_i(t))
\end{aligned}
$$

Table 4.3: Uniformization-based approaches (no constraints)

Citation	Complexity	
	Min.	Max.
[117]	$O(n\bar{k}^2)$	$O(n^2 m\bar{k}^2)$
[103]	$O(n\bar{k}^2)$	$O(n^2 a^m \bar{k}^2),\ a > 1$
[108]*	$O(n\bar{k}^2)$	$O(n^2 m\bar{k}^2/2)$
[341]	$O(nC^2 + nC(\bar{k} - C))$ iff $C < \bar{k}$	$O(n^2 mC^2 + n^2 C(\bar{k} - C))$ iff $C < \bar{k}$

*Computational cost for a single point of the probability distribution.

$$+ \sum_{j=0}^{n} \int_0^t c_{ij}(\tau) \sum_{k=0}^{m} \binom{m}{k} (r(i)\tau)^{m-k} M(j, t - \tau, k) d\tau.$$

4.7.2 Proof of theorem 4.1

We have to prove that $d(\phi(v), \phi(w)) \le \max_i \{Q_i(T)\} d(v, w), \forall v \in \mathcal{B}, \forall w \in \mathcal{B}$. By definition of $d(.,.)$ we have:

$$
\begin{aligned}
d(\phi(v), \phi(w)) &= \sup_{\substack{t \in [0,T] \\ y \in \mathcal{R}, i \in S}} \{|\phi(v(i, t, y)) - \phi(w(i, t, y))|\} \\
&= \sup_{\substack{t \in [0,T] \\ y \in \mathcal{R}, i \in S}} \{| \sum_{j=0}^{n} \int_0^t c_{ij}(\tau)(v(j, t - \tau, y - r(i)\tau) \\
&\qquad\qquad - w(j, t - \tau, y - r(i)\tau))d\tau|\} \\
&\le \sup_{\substack{t \in [0,T] \\ y \in \mathcal{R}, i \in S}} \{d(v, w)| \sum_{j=0}^{n} \int_0^t c_{ij}(\tau)d\tau|\} \\
&= d(\mathbf{v}, \mathbf{w}) \sup_{t \in [0,T]} \{|Q_i(t)|\}
\end{aligned}
$$

By definition, $Q_i(t)$ is an increasing function such that $Q_i(t) \leq 1$. If $Q_i(t) < 1 \ \forall i \in S$ and $\forall t \leq T$ then $\sup_t\{|Q_i(t)|\} = |Q_i(T)| < 1$, and the theorem is proved.

4.7.3   Proof of equation (4.13)

The recurrence-integral equation for $F(i, t, y|k)$ is as follows:

$$F(i, t, y|0) = \mathbf{1}(y - r(i)t)$$

$$F(i, t, y|k) = \sum_{j=0}^{n} p_{ij} \int_0^t \frac{k(t - \tau)^{k-1}}{t^k} F(j, t - \tau, y - r(i)\tau|k - 1)d\tau, \quad k > 0$$

The equation for $k = 0$ holds by definition. For $k > 0$, we proceed as follows. Let us denote by $T(\tau)$ the event designated as the first transition of the process $\mathbf{X}$ in the infinitesimal interval $(\tau, \tau + d\tau)$. By the theorem of total probability we have:

$$\begin{aligned} F(i, t, y|k) &\equiv \Pr\{Y_i(t) \leq y \mid k \text{ transitions}\} \\ &= \int_0^t \Pr\{Y_i(t) \leq y \mid T(\tau), k \text{ trans.}\}\Pr\{T(\tau) \mid k \text{ trans.}\}d\tau. \end{aligned}$$

The second term under the integral sign can be evaluated by using the well-known properties of the Poisson processes [86]. This term denotes the probability that the minimum of k random variables uniformly distributed in $[0, t]$ belongs to the interval $(\tau, \tau + d\tau)$, and thus:

$$\Pr\{T(\tau) \mid k \text{ transitions}\} = \frac{k(t - \tau)^{k-1}}{k}d\tau$$

By further conditioning on the state j towards which a transition at instant τ takes place, and by noting that $\Pr\{Y_i(t) \leq y \mid T(\tau), k \text{ transitions, first transition from } i \text{ to } j\}$ is equal to $\Pr\{r(i)\tau + Y_j(t - \tau) \leq y \mid k - 1 \text{ transitions}$ in $[\tau, t]\}$, which is again equal to $F(j, t - \tau, y - r(i)\tau \mid k - 1)$, we obtain the equation for $k > 0$.

Chapter 5

Markov-Reward Models and Hyperbolic Systems

Krishna R. Pattipati

Ranga Mallubhatla

V. Gopalakrishna

N. Viswanatham

We consider the problem of analyzing event-driven systems that evolve according to any one of a finite set of modes. The mode of evolution is controlled by a random mechanism that switches from one mode to another according to a continuous-time, homogeneous Markov chain. The reward rates in various modes of operation can be different, but are known constants. In this chapter, we show that the distribution of accumulated reward over a specified time interval, termed performability, is the solution of a system of either forward or adjoint linear hyperbolic partial differential equations (PDEs). We also show that the moments of performability satisfy a recursive set of ordinary differential equations (ODEs). Our approach provides a unified framework to interpret and extend existing numerical and

analytical solutions to the distribution of cumulative opera-
tional time and performability, as well as a vehicle to derive
asymptotic results, i.e., as the time interval tends to infinity.

5.1 Introduction

5.1.1 Motivation

CONSIDER the following four physical models:

1. A gracefully degradable system (e.g., a computer, communication network, a flight control system, a manufacturing system, a power network) operates in different modes of operation (also termed structure states) due to failure, repair, and reconfiguration of system resources. These events typically induce a change in the instantaneous benefit (reward, processing capacity, performance) accrued from the system.

2. A manoeuvring object moves in space with a constant velocity, until it changes its course; then the object changes its velocity, and again moves in space with a new constant velocity.

3. A signal propagates through a non-stationary medium, in which the index of refraction is changing at random.

4. A population of bacteria evolves in a randomly fluctuating environment.

These are all examples of event-driven systems, in which a system switches its "mode of evolution" or "dynamics of motion" due to causative events (as in Examples 1 and 2 above) or random changes in the environment (as in Examples 3 and 4). In recent years, these problems have been studied in mathematics, computer science and control theory under the rubric of random evolutions [150, 196, 228, 229, 373, 374], Markov reward models and performability (this book), jump linear systems [197, 301, 452, 475], and piecewise deterministic Markov processes [102]. It is the purpose of this chapter to interpret and extend existing numerical and analytical solutions in the study of Markov-reward models, via hyperbolic systems of partial differential equations (PDEs).

5.1.2   Markov-reward process

Assume that the event-driven system evolves according to any one of N structure states (modes), which we denote by $S = \{1, 2, \cdots, N\}$. The structure state evolution is controlled by a random mechanism $\{Z(t), t \geq 0\}$ that switches from one state to another according to a continuous-time, homogeneous Markov chain (CTMC) with state space S and infinitesimal generator matrix $\mathbf{Q} = [q_{ij}]$. It is well known [104] that if the CTMC is uniformized with an intensity parameter $\lambda \geq \max_i(-q_{ii})$, then $\mathbf{Q} = \lambda(\mathbf{P}-\mathbf{I})$, where $\mathbf{P} = [P_{ij}]$ is the matrix of transition probabilities associated with the uniformized chain and $\mathbf{I}$ is an $N \times N$ identity matrix.

Let $p_i(t) = \Pr\{Z(t) = i\}$ be the unconditional probability that the CTMC is in state i. Then, the $1 \times N$ row vector $\mathbf{p}(t) = [p_1(t), \cdots, p_N(t)]$ of state probabilities satisfies the ordinary differential equation (ODE):

$$\frac{d\mathbf{p}(t)}{dt} = \mathbf{p}(t)\mathbf{Q}, \tag{5.1}$$

with the initial condition $\mathbf{p}(0) = \mathbf{p}_0$. Equivalently,

$$\mathbf{p}(t) = \mathbf{p}_0 e^{\mathbf{Q}t} = \sum_{n=0}^{\infty} \Pi(n)e^{-\lambda t}\frac{(\lambda t)^n}{n!}, \tag{5.2}$$

where $\Pi(n+1) = \Pi(n)\mathbf{P}$ with the initial condition $\Pi(0) = \mathbf{p}_0$.

The steady-state probability vector π exists if the CTMC is irreducible and positive recurrent, and satisfies $\pi\mathbf{Q} = \mathbf{0}$; $\Sigma_{i \in S}\pi_i = 1$. That is, π is the *left* eigenvector of $\mathbf{Q}$ corresponding to the zero eigenvalue. In addition, since $q_{ii} = -\Sigma_{j=1, j\neq i}^{N} q_{ij}$, it is clear that $\mathbf{Q}\mathbf{e} = 0$, where $\mathbf{e}$ is an $N \times 1$ column vector defined by $\mathbf{e} = [1, 1, \cdots, 1]^T$ and the superscript ^{T} denotes the transpose operator. Thus, $\mathbf{e}$ is the *right* eigenvector corresponding to the zero eigenvalue.

Let $r_{Z(t)}$ be a nonnegative real-valued reward (performance) functional on the state-space S and let $Y(t)$ be an additive functional of the form:

$$Y(t) = \int_0^t r_{Z(\tau)}d\tau = \sum_{i=1}^{N} r_i\Gamma_i(t), \tag{5.3}$$

where $\Gamma_i(t)$ is the *occupation time* in state i defined by

$$\Gamma_i(t) = \int_0^t \delta_{iZ(\tau)}d\tau = \text{Lebesgue measure}\{\tau : 0 \le \tau \le t, Z(\tau) = i\} \quad (5.4)$$

and where $\delta_{iZ(t)}$ is the Kronecker delta function that has value 1 whenever $Z(t) = i$ and zero otherwise. Following Meyer [315], we define the performability of an event-driven system as the probability distribution function of $Y(t)$, denoted by $F_{Y(t)}(y,t) = \Pr\{Y(t) \le y\}$. The time average of $Y(t)$ and its distribution are denoted as:

$$W(t) = \frac{Y(t)}{t} = \frac{1}{t}\int_0^t r_{Z(\tau)}d\tau = \sum_{i=1}^N \frac{r_i\Gamma_i(t)}{t}, \quad (5.5)$$

and

$$F_{W(t)}(w,t) = \Pr\{W(t) \le w\} = \Pr\{Y(t) \le wt\} = F_{Y(t)}(wt,t). \quad (5.6)$$

In this chapter, we will investigate methods for computing $F_{Y(t)}(y,t)$, $F_{W(t)}(w,t)$, $F_{\Gamma_i(t)}(\gamma_i,t)$, $E\{Y^n(t)\}$ and $E\{W^n(t)\}$, as well as their asymptotic properties as $t \to \infty$.[1]

5.1.3 Scope and organization

We will show that the key to the evaluation of $F_{Y(t)}(y,t)$ is the computation of the joint distribution functions

$$\alpha_j(y,t) = \Pr\{Y(t) \le y, Z(t) = j\}, \quad j = 1, 2, \cdots, N, \quad (5.7)$$

and/or the conditional distribution functions

$$\beta_i(y,t) = \Pr\{Y(t) \le y | Z(0) = i\}, \quad i = 1, 2, \cdots, N. \quad (5.8)$$

---

[1]Although we are considering simplified Markov-reward models, the results of this chapter can be extended to significantly more general models. These include: (i) non-homogeneous Markov processes, where the generator matrix $\mathbf{Q}$ is a function of t [370, 371]; (ii) $r_{Z(t)}$ can be any real-valued functional with impulses at a finite set of points (equivalently, $Y(t)$ is a discontinuous additive functional) [374]; (iii) $r_{Z(t)}$ can be a function of $Y(t)$ and $Z(t)$ [102]; (iv) $\mathbf{r}_{Z(t)}$ and $\mathbf{Y}(t)$ can be vectors [471]; and (v) discrete-time Markov reward models [298, 166].

Define the $1 \times N$ row vector $\alpha(y, t) = [\alpha_1(y, t), \alpha_2(y, t), \cdots, \alpha_N(y, t)]$ of joint distributions of $(Y(t), Z(t))$ and the N-column vector

$$\beta(y, t) = [\beta_1(y, t), \beta_2(y, t), \cdots, \beta_N(y, t)]^T$$

of conditional distributions of $Y(t)$ given $Z(0)$. We will state in Section 5.2 that $\alpha(y, t)$ and $\beta(y, t)$ satisfy dual systems of linear, hyperbolic PDEs. Once $\alpha(y, t)$ and/or $\beta(y, t)$ are computed, $F_{Y(t)}(y, t)$ can be evaluated as:

$$\alpha(y, t)\mathbf{e} = \mathbf{p}_0\beta(y, t) = \int_0^y \alpha(y - x, t - \tau)\frac{\partial\beta(x, \tau)}{\partial x}dx, \forall \tau \in [0, t] \quad (5.9)$$

In Section 5.2, we will also demonstrate a characteristic feature of Markov-reward models that the nth moment of performability is a function of the $(n-1)$th conditional moments. To do this, we define the $1 \times N$ row vector $\mathbf{l}_n(t) = [l_{n1}(t), l_{n2}(t), \cdots, l_{nN}(t)]$ and

$$\mathbf{m}_n(t) = [m_{n1}(t), m_{n2}(t), \cdots, m_{nN}(t)]^T,$$

which is an N-column vector, such that

$$l_{nj}(t) = E\{Y^n(t), \delta_{jZ(t)}\}, \quad j = 1, 2, \cdots, N,$$

$$m_{ni}(t) = E\{Y^n(t)|Z(0) = i\}, \quad i = 1, 2, \cdots, N.$$

Specifically, we will show that $\mathbf{l}_n(t)$ and $\mathbf{m}_n(t)$ evolve according to a recursive set of ODEs. Once $\mathbf{l}_n(t)$ and/or $\mathbf{m}_n(t)$ are computed, the nth moments of $Y(t)$ and $W(t)$, denoted by $\overline{Y^n(t)}$ and $\overline{W^n(t)}$, are given, for all$k = 0, \cdots, n - 1$, by:

$$\overline{Y^n(t)} = t^n\overline{W^n(t)} = \mathbf{l}_n(t)\mathbf{e} = \mathbf{p}_0\mathbf{m}_n(t)$$

$$= n\binom{n-1}{k}\int_0^t \mathbf{l}_k(t - \tau)\mathbf{R}\mathbf{m}_{n-k-1}(\tau)d\tau. \quad (5.10)$$

The method of characteristics (MOC) used to solve the hyperbolic system of PDEs provides a uniformized version of the renewal equation associated with $\beta_i(y, t)$ (or equivalently $\alpha_j(y, t)$). This form of renewal equation enables us to show that the existing algorithms based on uniformization [104, 103] and partial fraction expansion [135] are analytic implementations of Picard's (fixed point) iteration used to solve the hyperbolic PDEs by the MOC. In addition, the hyperbolic systems can be used to study the asymptotic properties of $F_{Y(t)}(y,t)$, $F_{W(t)}(w,t)$, and $F_{\Gamma_i(t)}(\gamma_i,t)$ as $t \to \infty$. These are the subject of Section 5.3. Finally, in Section 5.4, we provide a summary and future research directions. The proofs of results are omitted for brevity; they may be found in [370, 371, 286, 179].

5.2 Duality in Markov-reward models

5.2.1 Dual systems of hyperbolic PDEs

Since $r_{Z(t)}$ is a function of $Z(t)$ only and since $\{Z(t)\}$ is homogeneous, the composite process $\{Y(t), Z(t), t \geq 0\}$ is a homogeneous Markov process. The evolution of the distribution of the composite process $\{Y(t), Z(t), t \geq 0\}$ is characterized by the following theorem.

Theorem 5.1 *The N-row vector of distributions $\alpha(y,t)$ satisfies:*

$$\frac{\partial \alpha(y,t)}{\partial t} + \frac{\partial \alpha(y,t)}{\partial y}\mathbf{R} = \alpha(y,t)\mathbf{Q} = \lambda\alpha(y,t)(\mathbf{P} - \mathbf{I}) \qquad (5.11)$$

where $\mathbf{R} = \mathrm{diag}(r_1, r_2, \cdots, r_N)$, $\alpha(0,t) = \mathbf{0}, t > 0$, $\alpha(y,0) = \mathbf{p}_0 U(y)$, and $U(y) = 1$ if $y \geq 0$ and zero otherwise.

Proof. See [371, 286]. □

Remarks:

1. The system of PDEs in (5.11) is hyperbolic, since the diagonal matrix $\mathbf{R}$ is *symmetric* with *real* eigenvalues $(r_1, r_2, \cdots, r_N)$ [149].

2. The partial derivatives in (5.11) should be interpreted in a generalized sense of differentiability (where the derivatives of unit-step functions are impulses) due to jump discontinuities in the initial conditions $\alpha(y,0) = \mathbf{p}_0 U(y)$. Indeed, the numerical difficulties associated with the evaluation of performability via finite difference methods or numerical method of lines [371, 179] stem from these discontinuities in the initial conditions, since they propagate over time and manifest themselves as jumps in $\alpha(y,t)$. We can partially overcome the numerical difficulties by approximating the unit

step functions $U(y)$ by a differentiable distribution function, such as $\alpha(y, 0) = [1 - exp(-y/\varepsilon)]\mathbf{p_0}$ for sufficiently small ε. An alternative method of addressing the differentiability issue that exploits the linearity of the PDEs in (5.11) is discussed in [371].

3. From the definition of $Y(t)$ and the non-negativity of $\{r_i\}$, it is clear that every sample path of $Y(t)$ is a continuous, non-negative and monotonically non-decreasing function of time. Therefore, $\alpha(y, t) = \mathbf{0}$, for $y < 0$. Consequently, the boundary condition is $\alpha(0, t) = \mathbf{0}, t \geq 0$.

4. As we noted earlier, $F_{Y(t)}(y, t) = \alpha(y, t)\mathbf{e}$. If we are given the initial probability distribution $\mathbf{p_0}$, then by computing $\alpha(y, t)$ for various values of t, we can evaluate the distribution of $Y(t)$ at various times. This is why (5.11) is called a *forward* PDE.

Often, for example [136, 104, 370, 371, 440, 439], we are required to compute the conditional density of $Y(t)$ given the initial structure state $Z(0)$. In this case, we consider the N-column vector of distributions $\beta(y, t)$, whose evolution is given by the following theorem.

Theorem 5.2 *The N-column vector of distributions $\beta(y, t)$ satisfies:*

$$\frac{\partial \beta(y, t)}{\partial t} + \mathbf{R}\frac{\partial \beta(y, t)}{\partial y} = \mathbf{Q}\beta(y, t) = \lambda(\mathbf{P} - \mathbf{I})\beta(y, t), \tag{5.12}$$

with the boundary and initial conditions: $\beta(0, t) = \mathbf{0}$ *for* $t > 0$ *and* $\beta(y, 0) = \mathbf{e}U(y)$.

Proof. See [371, 286]. □

Once we compute $\beta(y, t)$, we can determine the distribution of $Y(t)$ by using the fact that $F_{Y(t)}(y, t) = \mathbf{p_0}\beta(y, t)$. This is the preferred approach, if it is desired to compute $F_{Y(t)}(y, t)$ for various initial configurations *and* various times t.

Remarks:

1. The PDEs in (5.11) and (5.12) are said to be duals of each other (note the position of $\mathbf{Q}$ in (5.11) and (5.12)). This duality is pursued in greater detail in [371]. We call (5.12) an adjoint PDE.

2. The adjoint PDE in (5.12) has a major computational advantage over that in (5.11) for the homogeneous Markov model considered in this chapter: while the PDE in (5.11) must be solved *repeatedly* for various initial structure states, the PDE in (5.12) needs to be solved *once* for any initial structure state *and* any time t. Therefore, the adjoint PDE is superior to the forward PDE in the homogeneous case.

3. The conditional distribution of occupation time $\Gamma_i(t)$ given $Z(0)$ can easily be obtained from (5.12) by setting $\mathbf{R} = \mathbf{e}_i \mathbf{e}_i^T$, where $\mathbf{e}_i$ is an N-column vector with 1 in the ith place and zeros everywhere else (see Section 5.5).

5.2.2   Duality in the transform domain

The duality of (5.11) and (5.12) is transparent when we consider the Laplace transforms of $\alpha(y,t)$ and $\beta(y,t)$.

Corollary 5.1 *Suppose we associate the transform variables (μ, s) with (y, t), then the double Laplace transforms of $F_{Y(t)}(y,t)$, $\alpha(y,t)$ and $\beta(y,t)$, denoted by $F_y(\mu, s)$, $\alpha(\mu, s)$ and $\beta(\mu, s)$, respectively, are given by:*[2]

$$
\begin{aligned}
F_y(\mu, s) &= \frac{1}{\mu} \mathbf{p}_0 (sI + \mu \mathbf{R} - \mathbf{Q})^{-1} \mathbf{e} \\
&= \alpha(\mu, s)\mathbf{e} = \mathbf{p}_0 \beta(\mu, s) \qquad\qquad (5.13)
\end{aligned}
$$

$$
\alpha(\mu, s) = \frac{1}{\mu} \mathbf{p}_0 (sI + \mu \mathbf{R} - \mathbf{Q})^{-1} \qquad\qquad (5.14)
$$

$$
\beta(\mu, s) = \frac{1}{\mu} (sI + \mu \mathbf{R} - \mathbf{Q})^{-1} \mathbf{e} \qquad\qquad (5.15)
$$

The double Laplace transforms $F_y(\mu, s)$, $\alpha(\mu, s)$ and $\beta(\mu, s)$ can be readily inverted with respect to s to obtain $F_y(\mu, t)$, $\alpha(\mu, t)$ and $\beta(\mu, t)$.

---

[2]Note that there is a slight abuse of notation here: we are using the same variable to denote the transformed function as well as the original function. However, since we always use (μ, s) as arguments of transformed functions, there should not be any confusion. The only exception is the subscript y to denote that the distribution of y is being transformed, to distinguish it from w. This provides a uniform notation to denote partially transformed functions.

Corollary 5.2 $F_y(\mu, t)$, $\alpha(\mu, t)$ and $\beta(\mu, t)$ are given by:

$$
\begin{aligned}
F_y(\mu, t) &= \frac{1}{\mu}\mathbf{p}_0 e^{(\mathbf{Q}-\mu\mathbf{R})t}\mathbf{e} = \alpha(\mu, t)\mathbf{e} \\
&= \mathbf{p}_0\beta(\mu, t) = \mu\alpha(\mu, t - \tau)\beta(\mu, \tau), \ 0 \le \tau \le t \quad (5.16)
\end{aligned}
$$

$$
\alpha(\mu, t) = \frac{1}{\mu}\mathbf{p}_0 e^{(\mathbf{Q}-\mu\mathbf{R})t} \quad (5.17)
$$

and

$$
\beta(\mu, t) = \frac{1}{\mu}e^{(\mathbf{Q}-\mu\mathbf{R})t}\mathbf{e}. \quad (5.18)
$$

Note that the convolution integral in (5.9) readily follows from $F_y(\mu, t) = \mu\alpha(\mu, t - \tau)\beta(\mu, \tau)$. Now let us consider the distribution of time average, $F_{w(t)}(w, t)$. Since

$$
F_{w(t)}(w, t) = F_{Y(t)}(wt, t),
$$

we have the following corollary from the time scaling property of Laplace transforms.

Corollary 5.3 If we associate the Laplace transform variable σ with w, then $F_{w(t)}(\sigma, t)$ is given by:

$$
F_w(\sigma, t) = \frac{1}{t}F_y(\frac{\sigma}{t}, t) = \frac{1}{\sigma}\mathbf{p}_0 e^{(\mathbf{Q}t-\sigma\mathbf{R})}\mathbf{e} = \frac{1}{t}\alpha(\frac{\sigma}{t}, t)\mathbf{e} = \frac{1}{t}\mathbf{p}_0\beta(\frac{\sigma}{t}, t) \quad (5.19)
$$

Corollaries 5.2 and 5.3 have been used in [374] to study the asymptotic properties of $F_{w(t)}(w, t)$, $F_{Y(t)}(y, t)$ and $F_{\Gamma_i(t)}(\gamma_i, t)$.

5.2.3 Dual systems of moment recursions

A characteristic feature of Markov reward models is that the nth moment of performability can be expressed in terms of $(n-1)$th moments. Using the dual system of PDEs in (5.11) and (5.12), we can readily compute the nth moment of the composite process $\{Y(t), Z(t), t \ge 0\}$ and of the conditional process $\{Y(t), t \ge 0 | Z(0)\}$.

Corollary 5.4 The nth moment vector of the composite process $\{Y(t), Z(t), t \ge 0\}$ evolves according to the ODE:

$$
\frac{d\mathbf{l}_n(t)}{dt} = \mathbf{l}_n(t)\mathbf{Q} + n\mathbf{l}_{n-1}(t)\mathbf{R}, \quad (5.20)
$$

where $l_0(t) = \mathbf{p}(t)$ and $l_n(0) = 0$, $n \geq 1$. Furthermore, the Laplace transform of $l_n(t)$, denoted by $l_n(s)$, is given by:

$$l_n(s) = n!\mathbf{p}_0(s\mathbf{I} - \mathbf{Q})^{-1}[\mathbf{R}(s\mathbf{I} - \mathbf{Q})^{-1}]^n = n!\mathbf{p}_0[(s\mathbf{I} - \mathbf{Q})^{-1}\mathbf{R}]^n(s\mathbf{I} - \mathbf{Q})^{-1}. \tag{5.21}$$

Proof. See [371, 286]. □

Corollary 5.5 *The nth conditional moment vector* $\mathbf{m}_n(t)$ *evolves according to the equation:*

$$\frac{d\mathbf{m}_n(t)}{dt} = \mathbf{Q}\mathbf{m}_n(t) + n\mathbf{R}\mathbf{m}_{n-1}(t), \tag{5.22}$$

where $\mathbf{m}_0(t) = \mathbf{e}$ *and* $\mathbf{m}_n(0) = 0$, $n \geq 1$. *In addition, the Laplace transform of* $\mathbf{m}_n(t)$, *denoted by* $\mathbf{m}_n(s)$, *is given by:*

$$\mathbf{m}_n(s) = n![(s\mathbf{I} - \mathbf{Q})^{-1}\mathbf{R}]^n \frac{\mathbf{e}}{s} = n![(s\mathbf{I} - \mathbf{Q})^{-1}\mathbf{R}]^n(s\mathbf{I} - \mathbf{Q})^{-1}\mathbf{e}, \tag{5.23}$$

the last equality stemming from $(s\mathbf{I} - \mathbf{Q})^{-1}\mathbf{e} = \dfrac{\mathbf{e}}{s}$.

Proof. See [371, 286]. □

Evidently, the Laplace transform of the nth moment of $Y(t)$ is (for $0 \leq k \leq n - 1$):

$$\mathcal{L}\left\{\overline{Y^n(t)}\right\} = l_n(s)\mathbf{e} = \mathbf{p}_0\mathbf{m}_n(s) = n\binom{n-1}{k}l_k(s)\mathbf{R}\mathbf{m}_{n-k-1}(s). \tag{5.24}$$

From (5.24), the convolution relation in (5.10) follows immediately.

In many practical situations [192, 471, 179], we are primarily interested in the first two moments of performability. Using (5.2) and (5.22–5.24), it is straightforward to show that

$$\begin{aligned}
\overline{Y(t)} &= \mathbf{p}_0\int_0^t e^{\mathbf{Q}\tau}\mathbf{r}d\tau \\
&= \frac{1}{\lambda}\sum_{n=0}^{\infty}\frac{e^{-\lambda t}(\lambda t)^{n+1}}{(n+1)!}\left[\sum_{k=0}^{n}\Pi(k)\right]\mathbf{r} \\
&= \frac{1}{\lambda}\sum_{n=0}^{\infty}a_n\left[\sum_{k=0}^{n}\Pi(k)\right]\mathbf{r}, \tag{5.25}
\end{aligned}$$

where $\mathbf{r} = [r_1, r_2, \cdots, r_N]^T$ is an N-column vector and

$$a_n = \frac{e^{-\lambda t}(\lambda t)^{n+1}}{(n+1)!} = \frac{\lambda t}{(n+1)}a_{n-1}, \quad a_0 = \lambda t e^{-\lambda t}. \tag{5.26}$$

Also,

$$\overline{Y(t)} = \mathbf{p}_0 \mathbf{m}_1(t) = \mathbf{p}_0 \sum_{n=0}^{\infty} a_n \mathbf{b}_n, \tag{5.27}$$

where $\mathbf{b}_n$ is an N-column vector given by the recursion

$$\mathbf{b}_n = \frac{1}{\lambda} \sum_{k=0}^{n} \mathbf{P}^k \mathbf{r} = \mathbf{P}\mathbf{b}_{n-1} + \frac{\mathbf{r}}{\lambda}, \quad \mathbf{b}_0 = \frac{\mathbf{r}}{\lambda}. \tag{5.28}$$

Similarly, the second moment is given by

$$\begin{aligned}
\overline{Y^2(t)} &= 2\mathbf{p}_0 \int_0^t e^{\mathbf{Q}(t-\tau)} \mathbf{R}\mathbf{m}_1(\tau)d\tau \\
&= \frac{2}{\lambda} \sum_{n=0}^{\infty} \sum_{k=0}^{\infty} a_{k+n+1} \Pi(n) \mathbf{R}\mathbf{b}_k.
\end{aligned} \tag{5.29}$$

A recursive finite sum approximation of $\overline{Y(t)}$ and $\overline{Y^2(t)}$ to any desired accuracy can easily be derived from (5.27) and (5.29) [179].

5.3 Method of characteristics and uniformization

5.3.1 Uniformized renewal equation and the MOC

Hyperbolic PDEs arise in many transport phenomena (neutron diffusion, heat transfer, radiative transfer), wave mechanics, gas dynamics, vibrations and others [454, 393]. In order to motivate the method of characteristics, consider the following scalar PDE:

$$\frac{\partial \beta_i(y, t)}{\partial t} + r_i \frac{\partial \beta_i(y, t)}{\partial y} = 0, \quad r_i \geq 0. \tag{5.30}$$

This equation is referred to as the advection (convection) equation. It describes a wave moving in the y direction with a speed of r_i units per unit of time. If the initial conditions are given as $\beta_i(y,0) = U(y)$, then the double Laplace transform $\beta_i(\mu, s) = \frac{1}{\mu(s+\mu r_i)}$. Consequently, $\beta_i(y,t) = U(y - r_i t)$. Thus, the value of β_i at any point (y,t) is determined by the value at the point $(y - r_i t, 0)$. In other words, β_i is constant along the lines $y - r_i t = $ constant, and these are called the characteristic curves of (5.30). Now consider a slightly more general scalar PDE:

$$\frac{\partial \beta_i(y,t)}{\partial t} + r_i \frac{\partial \beta_i(y,t)}{\partial y} + \lambda \beta_i(y,t) = 0, \quad \lambda \geq 0. \tag{5.31}$$

The solution of (5.31) with the initial condition $\beta_i(y,0) = U(y)$ is $\beta_i(y,t) = e^{-\lambda t}U(y - r_i t)$. This equation describes a *dissipative* wave in the y direction with a speed of r_i. A more general class of equations is the coupled set of PDEs in (5.12):

$$\frac{\partial \beta_i(y,t)}{\partial t} + r_i \frac{\partial \beta_i(y,t)}{\partial y} + \lambda \beta_i(y,t) = \lambda \sum_{j=1}^{N} P_{ij} \beta_j(y,t), \quad i = 1, 2, \cdots, N, \tag{5.32}$$

or in the transformed domain:

$$\beta_i(\mu, s) = \frac{1}{\mu(s + \lambda + \mu r_i)} + \lambda \sum_{j=1}^{N} P_{ij} \frac{\beta_j(\mu, s)}{(s + \lambda + \mu r_i)}. \tag{5.33}$$

Inverting (5.33) with respect to s first and then with respect to μ, we obtain the *uniformized* renewal equation:

$$\beta_i(y,t) = e^{-\lambda t}U(y - r_i t) + \lambda \sum_{j=1}^{N} P_{ij} \int_0^t e^{-\lambda \tau} \beta_j(y - r_i \tau, t - \tau)d\tau. \tag{5.34}$$

Relations (5.32–5.34) are the keys to establishing connections with the previous numerical approaches to performability evaluation based on uniformization.

Remarks:

1. There exists another useful interpretation of the uniformized renewal equation. Suppose we employ a scalar transformation:

$$\mathbf{f}(y,t) = e^{\lambda t}\beta(y,t), \tag{5.35}$$

then it is straightforward to show that

$$\frac{\partial \mathbf{f}(y,t)}{\partial t} + \mathbf{R}\frac{\partial \mathbf{f}(y,t)}{\partial t} = (\mathbf{Q} + \lambda \mathbf{I})\mathbf{f}(y,t) = \lambda \mathbf{P}\mathbf{f}(y,t). \qquad (5.36)$$

Using (5.30), the solution of (5.36) can be shown to be:

$$f_i(y,t) = U(y - r_i t) + \lambda \sum_{j=1}^{N} P_{ij} \int_0^t f_j(y - r_i \tau, t - \tau)d\tau. \qquad (5.37)$$

Using the facts that $\beta_i(y,t) = e^{-\lambda t} f_i(y,t)$ and that $f_j(y - r_i \tau, t - \tau) = e^{\lambda(t-\tau)}\beta_j(y - r_i \tau, t - \tau)$, we obtain the renewal equation in (5.34).

2. Another way of deriving the renewal equation in (5.34) is to note the well-known matrix identity [178]:

$$e^{(\mathbf{Q}-\mu\mathbf{R})t} = e^{(-(\mu\mathbf{R}+\lambda\mathbf{I})+\lambda\mathbf{P})t},$$

so that

$$e^{(\mathbf{Q}-\mu\mathbf{R})t} = e^{-(\mu\mathbf{R}+\lambda\mathbf{I})t} + \lambda \int_0^t e^{-(\mu\mathbf{R}+\lambda\mathbf{I})(t-\tau)}\mathbf{P}e^{(\mathbf{Q}-\mu\mathbf{R})\tau}d\tau. \qquad (5.38)$$

After inverting (5.38) with respect to μ and noting that $\beta(\mu,t) = \frac{1}{\mu}e^{(\mathbf{Q}-\mu\mathbf{R})t}\mathbf{e}$, we obtain (5.34).

3. Since $y \geq 0$, the upper limit in the integration in (5.34) can be replaced by $\min(t, y/r_i)$. In addition, the range of y can be limited to $r_{\min}t \leq y \leq r_{\max}t$, where

$$r_{\min} = \min_{1 \leq i \leq N} r_i \qquad \text{and} \qquad r_{\max} = \max_{1 \leq i \leq N} r_i.$$

This is because $\beta_i(y,t) = 0$ for $y < r_{\min}t$ and $\beta_i(y,t) = 1$ for $y \geq r_{\max}t$.

5.3.2 Uniformization as the analytic solution of an infinite set of scalar PDEs

The uniformization-based methods for computing performability distribution are based on the following idea. Let, for $i = 1, \cdots, N$,

$$\beta_i(y,t;n) = \Pr\{Y(t) \leq y; \ n \text{ transitions in } (0,t)|Z(0) = i\}. \qquad (5.39)$$

Then $\beta(y,t)$ and $\beta(\mu,s)$ are given by

$$\beta(y,t) = \sum_{n=0}^{\infty} \beta(y,t;n) \tag{5.40}$$

$$\beta(\mu,s) = \sum_{n=0}^{\infty} \beta(\mu,s;n). \tag{5.41}$$

Now consider the double Laplace transform equation in (5.33). It can be rewritten in vector-matrix notation as:

$$\beta(\mu,s) = \mathbf{D}(\mu,s)\beta(\mu,0) + \lambda\mathbf{D}(\mu,s)\mathbf{P}\beta(\mu,s), \tag{5.42}$$

where

$$\mathbf{D}(\mu,s) = \operatorname{diag}\left[\frac{1}{(s+\lambda+\mu r_1)}, \frac{1}{(s+\lambda+\mu r_2)}, \cdots, \frac{1}{(s+\lambda+\mu r_N)}\right] \tag{5.43}$$

and $\beta(\mu,0) = \frac{1}{\mu}\mathbf{e}$. Thus,

$$\begin{aligned}
\beta(\mu,s) &= [\mathbf{I} - \lambda\mathbf{D}(\mu,s)\mathbf{P}]^{-1}\mathbf{D}(\mu,s)\beta(\mu,0) \\
&= \sum_{n=0}^{\infty}[\lambda\mathbf{D}(\mu,s)\mathbf{P}]^n\mathbf{D}(\mu,s)\beta(\mu,0).
\end{aligned} \tag{5.44}$$

Comparing (5.41) and (5.44), we have:[3]

$$\beta(\mu,s;n) = [\lambda\mathbf{D}(\mu,s)\mathbf{P}]^n\mathbf{D}(\mu,s)\beta(\mu,0) = \lambda\mathbf{D}(\mu,s)\mathbf{P}\beta(\mu,s;n-1), \tag{5.45}$$

where $\beta(\mu,s;0) = \mathbf{D}(\mu,s)\beta(\mu,0)$ implying that $\beta_i(y,t;0) = e^{-\lambda t}U(y - r_i t), i = 1, 2, \cdots, N$. We have, from (5.45), the *convolution* integral linking the conditional performability with n transitions and that with $n-1$ transitions and $i = 1, 2, \cdots, N$:

$$\begin{aligned}
\beta_i(y,t;n) &= \lambda\sum_{j=1}^{N} P_{ij}\int_0^t e^{-\lambda\tau}\beta_j(y - r_i\tau, t - \tau; n-1)d\tau \\
&= \lambda\sum_{j=1}^{N} P_{ij}\int_0^{\Delta t} e^{-\lambda\tau}\beta_j(y - r_i\tau, t - \tau; n-1)d\tau \\
&\quad + e^{-\lambda\Delta t}\beta_i(y - r_i\Delta t, t - \Delta t; n), \quad 0 \le \Delta t \le t. \tag{5.46}
\end{aligned}$$

---

[3]A formal proof of equivalence is straightforward.

The recursion in (5.46) is initialized with $\beta_i(y, t; 0) = e^{-\lambda t}U(y - r_i t)$ and $\beta_i(y, 0; n) = 0$ for $n \geq 1$. Indeed, $\beta(y, t; n)$ satisfy the *infinite set of scalar PDEs* (for $i = 1, 2, \cdots, N$ and $n \geq 1$):

$$\frac{\partial \beta_i(y, t; n)}{\partial t} + r_i \frac{\partial \beta_i(y, t; n)}{\partial y} + \lambda \beta_i(y, t; n) = \lambda \sum_{j=1}^{N} P_{ij} \beta_j(y, t; n - 1). \quad (5.47)$$

Thus, uniformization corresponds to the solution of *decoupled* PDEs in (5.47) at each stage n. In order to obtain the structure of the solution of $\beta_i(\mu, s; n)$, let $r_i \in \{\rho_1, \rho_2, \cdots, \rho_L\}$, where $L \leq N$ is the number of distinct rewards of the event-driven system. That is, $r_i = \rho_l$ if the reward in structure state i corresponds to the lth reward. Then, from (5.45), it is straightforward to see that:

$$\beta(\mu, s; n) = \frac{1}{\mu} D(\mu, s) \sum_{k: \sum_{l=1}^{L} k_l = n} \frac{c_n(k)}{\prod_{l=1}^{L}(s + \lambda + \mu \rho_l)^{k_l}}, \quad (5.48)$$

where $k = (k_1, k_2, \cdots, k_L)$ and k_l denotes the number of transitions of the system to states with reward rate ρ_l out of a total of n transitions during $(0, t)$. The coefficient vector $c_n(k) = [c_{n1}(k), c_{n2}(k), \cdots, c_{nN}(k)]^T$ satisfies the recursion:

$$c_{ni}(k) = \sum_{j=1, r_j=\rho_l}^{N} P_{ij} c_{n-1,j}(k - e_l), \quad c_{0i}(0) = 1. \quad (5.49)$$

This recursion, which is essentially equivalent to the method of De Souza e Silva and Gail [104, 103], suffers from *exponential* increase in computation with the number of distinct rewards, L. However, when computing the distribution of cumulative operational time (i.e., $L = 2$, $\rho_1 = 1$, and $\rho_2 = 0$), the method leads to elegant, stable recursions as shown in the appendix in Section 5.5.

A clever idea, due to Donatiello and Grassi [135], involves replacing (5.48) by its partial fraction expansion. Indeed, we can show that:

$$\beta_i(\mu, s; n) = \frac{a_{in}}{(s + \lambda + \mu r_i)^{n+1}} + \sum_{k=1}^{n} \sum_{l=1}^{L} \frac{b_{in}(k, l)}{\mu^{n-k+1}(s + \lambda + \mu \rho_l)^k}. \quad (5.50)$$

Using (5.45), one can derive recursive expressions for the coefficients a_{in} and $b_{in}(k, l)$ that are identical to those in Donatiello and Grassi [135], leading to a polynomial algorithm for performability evaluation. However, preliminary computational experiments on a two-state availability model with repair have shown the algorithm to be *numerically unstable* [179]. This is because the coefficients $b_{in}(k, l)$ can have positive and negative signs leading to severe numerical cancellation errors. Our limited computational experiments have shown that (5.46) with Δt equal to the time discretization step and the fixed point iteration, discussed below, are reliable and efficient methods for computing performability.

5.3.3 Successive approximation (Picard's method)

The fixed point iteration method used to solve hyperbolic PDEs in (5.12) over a time interval $[0, t_f]$ works as follows.

Algorithm FixedPoint:
  Initialize:
    $\beta_i^{(0)}(y, t) = e^{-\lambda t} U(y - r_i t)$, $0 \le t \le t_f$ and $0 \le y \le r_{\max} t_f$
    $\beta_i^{(0)}(y, t) = e^{-\lambda t}$ for $y > r_{\max} t_f$
  Iteration count $n = 0$
  ϵ = desired tolerance
  do until $\max_{1 \le i \le N} |\beta_i^{(n+1)} - \beta_i^{(n)}| < \epsilon$, for all y and t
    $\beta_i^{(n+1)}(y, t) = e^{-\lambda t} U(y - r_i t) + \lambda \sum_{j=1}^{N} P_{ij} \int_0^t e^{-\lambda \tau} \beta_j^{(n)}(y - r_i \tau, t - \tau) d\tau$
    $n = n + 1$
  end do

From (5.46), it is clear that

$$
\begin{aligned}
\beta_i^{(0)}(y, t) &= \beta_i(y, t; 0) \\
\beta_i^{(1)}(y, t) &= \beta_i(y, t; 0) + \beta_i(y, t; 1) \\
\beta_i^{(2)}(y, t) &= \beta_i(y, t; 0) + \beta_i(y, t; 1) + \beta_i(y, t; 2) \\
&\ \vdots \\
\beta_i^{(n)}(y, t) &= \sum_{k=0}^{n} \beta_i(y, t; k) \qquad\qquad (5.51)
\end{aligned}
$$

Thus, fixed point iteration corresponds to successively evaluating the partial sums $\sum_{k=0}^{n} \beta_i(y, t; k)$, and the uniformization method of [104] can be viewed as an analytic implementation of the convolution integral associated with the Picard's iteration.

Remarks:

1. The existence and uniqueness of solutions to the hyperbolic PDEs and of the uniform convergence of the sequence $\beta_i^{(n)}(y,t)$ follow from the fixed point iteration. The uniform convergence of the sequence $\{\beta_i^{(n)}(y,t)\}$ is equivalent to the uniform convergence of the series $\beta_i^{(0)}(y,t) + \sum_{n=0}^{\infty} \left[\beta_i^{(n+1)}(y,t) - \beta_i^{(n)}(y,t) \right] = \sum_{n=0}^{\infty} \beta_i(y,t;n)$. Since the functions $\beta_i^{(0)}(y,t)$ and $\beta_i^{(1)}(y,t)$ are differentiable (in a generalized sense) in the closed region $\overline{G}$ bounded by $[0, t_f]$ and the characteristics $y = r_{\min} t_f$ and $y = r_{\max} t_f$, they are bounded in $\overline{G}$. Since $\max_i |\beta_i^{(0)}(y,t)| \le 1$, we have $|\beta_i^{(1)}(y,t) - \beta_i^{(0)}(y,t)| \le \lambda t$ for all i and $(y,t) \in \overline{G}$.

   Now consider

$$|\beta_i^{(2)}(y,t) - \beta_i^{(1)}(y,t)| \le \lambda \int_0^t \sum_{j=1}^N P_{ij} |\beta_j^{(1)}(y,t) - \beta_j^{(0)}(y,t)| dt$$

$$\le \frac{(\lambda t)^2}{2!}$$

$$\vdots$$

$$|\beta_i^{(n+1)}(y,t) - \beta_i^{(n)}(y,t)| \le \frac{(\lambda t)^{n+1}}{n+1!} \le \frac{(\lambda t_f)^{n+1}}{n+1!}, \tag{5.52}$$

   since $t \le t_f$. Since the series $\frac{(\lambda t_f)^n}{n!}$ converges, it follows that the series

$$\beta_i^{(0)}(y,t) + \sum_{n=0}^{\infty} \left[\beta_i^{(n+1)}(y,t) - \beta_i^{(n)}(y,t) \right]$$

   converges. This establishes the existence of the solutions. The uniqueness of the solution follows a similar approach [372].

2. The fixed point iteration scheme has been found to work well in practice. The step sizes Δy and Δt should be picked to satisfy the Courant–Freiderich–Levy (CFL) condition [454, 393]:

$$\left| \frac{r_{\max} \Delta t}{\Delta y} \right| \le 1. \tag{5.53}$$

3. The question of which integration scheme to use arises with regard to (5.34) and (5.46). Adams–Moulton formulas [69] work well, especially for small Δy and Δt.

5.3.4   Asymptotic properties

Consider the conditional mean of $W(t)$ given $Z(0) = i$ as $t \to \infty$:

$$
\lim_{t\to\infty} E\{W(t)|Z(0) = i\} = \lim_{t\to\infty} \frac{1}{t} \int_0^t E\left\{r_{Z(\tau)}|Z(0) = i\right\} d\tau
$$

$$
= \mathbf{e}_i^T \lim_{t\to\infty} \left[\frac{1}{t} \int_0^t e^{\mathbf{Q}\tau} d\tau\right] \mathbf{r}. \tag{5.54}
$$

Since $e^{\mathbf{Q}\tau} = \mathbf{e}\pi$ as $\tau \to \infty$, we have

$$
\lim_{t\to\infty} E\{W(t)|Z(0) = i\} = \pi\mathbf{r} = \bar{r} = \overline{W}(\infty). \tag{5.55}
$$

The weak law of large numbers states that $\lim_{t\to\infty} W(t) = \bar{r}$ in the sense of convergence in probability. The following theorem makes this statement precise.

Theorem 5.3 *Assume that the CTMC $\{Z(t)\}$ is irreducible and positive recurrent. Then*

$$
\lim_{t\to\infty} \mathbf{P}\left\{W(t) \le w|Z(0) = i\right\} = U(w - \bar{r}) \tag{5.56}
$$

Proof. Pinsky [374] proves this theorem by using the asymptotic expansion of $e^{(\mathbf{Q}t - \sigma\mathbf{R})}$ in (5.19). Here we provide an alternative proof via moment recursions (see Section 5.2).

Let $\{\lambda_k, \xi_k, \phi_k; 1 \le k \le N\}$ be the sets of eigenvalues, the right eigenvectors (N column vectors) and left eigenvectors (N row vectors) of $\mathbf{Q}$, respctively, with the association that $\lambda_1 = 0$, $\xi_1 = \mathbf{e}$, and $\phi_1 = \pi$. The first conditional moment vector is

$$
\lim_{t\to\infty} \mathbf{m}_1(t) = \lim_{t\to\infty} \int_0^t e^{\mathbf{Q}\tau} \mathbf{r} d\tau = \lim_{t\to\infty}[\bar{r}\mathbf{e}t + \mathbf{H}(t)\mathbf{r}], \tag{5.57}
$$

where

$$
\mathbf{H}(t) = \int_0^t (e^{\mathbf{Q}\tau} - \mathbf{e}\pi) d\tau = \sum_{k=2}^N \frac{\xi_k \phi_k (e^{\lambda_k t} - 1)}{\lambda_k}. \tag{5.58}
$$

Note that as $t \to \infty$, $\mathbf{H}(\infty)$ is a constant matrix given by

$$\mathbf{H}(\infty) = \int_0^\infty (e^{\mathbf{Q}\tau} - \mathbf{e}\pi)d\tau = -\sum_{k=2}^N \frac{\xi_k \phi_k}{\lambda_k}. \tag{5.59}$$

Similarly,

$$
\begin{aligned}
\lim_{t\to\infty} m_2(t) &= 2 \lim_{t\to\infty} \int_0^t e^{\mathbf{Q}(t-\tau)}\mathbf{R}\mathbf{m}_1(\tau)d\tau \\
&= 2 \lim_{t\to\infty} \int_0^t e^{\mathbf{Q}(t-\tau)}\mathbf{R}[\bar{r}\mathbf{e}\tau + \mathbf{H}(\tau)\mathbf{r}]d\tau \\
&= 2 \lim_{t\to\infty} \int_0^t [\mathbf{e}\pi + \sum_{k=2}^N e^{\lambda_k(t-\tau)}\xi_k\phi_k]\mathbf{R}[\bar{r}\mathbf{e}\tau + \mathbf{H}(\tau)\mathbf{r}]d\tau \\
&\approx (\bar{r}t)^2\mathbf{e} + 2\pi\mathbf{R}\mathbf{H}(\infty)\mathbf{r}te + 2\bar{r}H(\infty)\mathbf{r}t + O(1). \tag{5.60}
\end{aligned}
$$

Thus, $\lim_{t\to\infty} m_2(t) = (\bar{r}t)^2\mathbf{e} + O(t)$. In general,

$$\lim_{t\to\infty} m_n(t) = (\bar{r}t)^n\mathbf{e} + O(t^{n-1}). \tag{5.61}$$

This implies that

$$\lim_{t\to\infty} E\{W^n(t)|Z(0) = i\} = \bar{r}^n = \overline{W^n(\infty)}. \tag{5.62}$$

Noting that

$$\lim_{t\to\infty} E\left\{e^{-\sigma W(t)}|Z(0) = i\right\} = \lim_{t\to\infty} \sum_{n=0}^\infty \frac{E\{W^n(t)\}(-\sigma)^n}{n!} = e^{-\bar{r}\sigma}, \tag{5.63}$$

the result in (5.56) follows immediately. □

Remarks:

1. It is clear from (5.57) and (5.60) that the aysmptotic variance of $Y(t)$ given $Z(0) = i$, denoted by $\sigma_i^2(t)$, is given by

$$
\begin{aligned}
\lim_{t\to\infty} \sigma_i^2 &= \lim_{t\to\infty} m_{2i}(t) - m_{1i}^2(t) \\
&= 2\pi\mathbf{R}\mathbf{H}(\infty)\mathbf{r}t = 2\left(\sum_{k=1}^N \sum_{l=1}^N H_{kl}(\infty)\pi_k r_k r_l\right) t \\
&= vt, \quad i = 1, 2, \cdots, N. \tag{5.64}
\end{aligned}
$$

By the central limit theorem [374], this implies that $\left(\frac{Y(t)-\bar{r}t}{\sqrt{t}}\right)$ is a normal random variable with mean zero and variance v, irrespective of the initial state $Z(0)$. That is:

$$\lim_{t\to\infty} F_{Y(t)}(y\sqrt{t} + \bar{r}t, t) = \int_{-\infty}^{y/\sqrt{v}} \frac{e^{-u^2/2}}{\sqrt{2\pi}} du. \qquad (5.65)$$

2. The normalized occupation time $\frac{(\Gamma_i(t)-\pi_i t)}{\sqrt{t}}$ is asymptotically normal, with mean zero and variance $(2\pi_i H_{ii}(\infty))$.

3. Using the generalized results of performability for vector $\mathbf{Y}(t)$ in [471], the joint distribution of the random vector $(\frac{\Gamma_1(t)-\pi_1 t}{\sqrt{t}}, \cdots, \frac{\Gamma_N(t)-\pi_N t}{\sqrt{t}})$ can be shown to be a multivariate normal distribution with mean vector zero and covariance matrix $\Sigma = \{\Sigma_{ij}\}$, where

$$\Sigma_{ij} = \pi_i H_{ij}(\infty) + \pi_j H_{ji}(\infty). \qquad (5.66)$$

4. The matrix $\mathbf{H}(\infty)$ can also be obtained from the following equations [374]:

$$\mathbf{H}(\infty)\mathbf{e} = \mathbf{0}, \qquad (5.67)$$

$$\mathbf{Q}\mathbf{H}(\infty) = \mathbf{H}(\infty)\mathbf{Q} = -\mathbf{I} + \mathbf{e}\pi. \qquad (5.68)$$

Therefore, if we let $\mathbf{h}_i^T$ denote the ith row of $\mathbf{H}(\infty)$, then

$$\begin{pmatrix} \mathbf{e}^T \\ \mathbf{Q}^T \end{pmatrix} \mathbf{h}_i = \begin{pmatrix} 0 \\ -\mathbf{e}_i + \pi^T \end{pmatrix}, \quad i = 1, 2, \cdots, N. \qquad (5.69)$$

5.4   Conclusions and future directions

In this chapter, we have demonstrated that a dual system of hyperbolic PDEs and ODE moment recursions provide a natural framework to study Markov-reward processes associated with event-driven systems. This framework enables us to interpret the computational methods based on uniformization as analytic implementations of fixed point iteration associated with the method of characteristics. Based on limited computational experimentation to date, MOC employing numerical integration appears to be one of the few reliable and efficient methods to compute performability. More significantly, the fixed point iteration does not suffer either from exponential computational complexity associated with uniformization or from the numerical instabilities associated with the partial fraction expansion approach. In addition, the connection between fixed point iteration and uniformization has enabled us to derive an apparently new and stable method for computing the distribution of cumulative operational time. Finally, the moment recursions enable us to characterize the asymptotic distribution of performability and time-averaged reward.

There exist numerous unresolved research issues related to Markov-reward models:

- *Computational methods:* Although MOC appears to be the method of choice for performability evaluation, a thorough comparative analysis of various computational algorithms is warranted. Some techniques that deserve evaluation include: (a) parallel Fast Fourier Transform (FFT) based methods to solve decoupled PDEs in (5.47), since the key operation is a convolution integral in (5.46); (b) fixed point iteration with sophisticated integration schemes, e.g., higher order Adams–Moulton, Gauss–Kronrod quadrature rules; [250]; (c) the third–order accurate explicit finite-difference method of Warming–Kutler–Lomax (WKL) [473], (d) the numerical method of lines [371]; (e) Petrov–Galerkin methods employing Leguerre, Gauss–Jacobi, or Chebyshev polynomials, as basis functions [454]; and (f) Pade approximations to compute $e^{(\mathbf{Q} - \mu \mathbf{R})t}$ followed by numerical inversion of the Laplace transform with respect to μ.

- *More general models:* As stated in Section 5.1, the theory of this chapter can be extended to significantly more general models involving nonhomogeneous CTMCs, semi-Markov processes, discrete-time Markov chains, impulsive and random reward structures and vector rewards. Efficient computational algorithms for these general models remain to be developed.

- *Design and control issues:* The moment recursions and the asymptotic results should aid in formulating *tractable* design and control problems associated with event-driven systems (e.g., optimal design of reconfiguration/recovery/repair rules). In this vein, the well developed theory of controlled Markov chains, for example [35], should prove useful.

Acknowledgements

The authors would like to thank Vijaya Raghavan of the University of Connecticut for his help in the preparation of this chapter.

5.5   Appendix: Distribution of cumulative operational time

Let S_0 and S_f denote the sets of operational and nonoperational structure states of the event-driven system such that $S_0 \cup S_f = S$. The analysis below can be adapted to compute the conditional distribution of occupation times in a state i by letting $S_0 = \{i\}$ and $S_f = S \setminus \{i\}$. From (5.45) it follows that

$$\beta(\mu, s; n) = \frac{1}{\mu} D(\mu, s) \sum_{k=0}^{n} \mathbf{c}_n(k) \left(\frac{\lambda}{s + \lambda + \mu} \right)^k \left(\frac{\lambda}{s + \lambda} \right)^{n-k}, \quad \mathbf{c}_0(0) = \mathbf{e}, \tag{5.70}$$

where each term in the summation has the interpretation that the system transitions to *up* states k times and to *down* states $(n - k)$ times. For each component, we have:

$$\beta_i(\mu, s; n) = \begin{cases} \frac{1}{\mu \lambda} \sum_{k=0}^{n} c_{ni}(k) \left(\frac{\lambda}{s+\lambda+\mu} \right)^{k+1} \left(\frac{\lambda}{s+\lambda} \right)^{n-k}, & i \in S_0, \\ \frac{1}{\mu \lambda} \sum_{k=0}^{n} c_{ni}(k) \left(\frac{\lambda}{s+\lambda+\mu} \right)^{k} \left(\frac{\lambda}{s+\lambda} \right)^{n-k+1}, & i \in S_f. \end{cases} \tag{5.71}$$

Let us consider $\beta_i(\mu, s; n)$ corresponding to an operational state $i \in S_0$. Inverting with respect to μ we obtain:

$$\beta_i(y, s; n) = \sum_{k=0}^{n} c_{ni}(k) \left(\int_0^y e^{-(s+\lambda)y} \frac{(\lambda y)^k}{k!} dy \right) \left(\frac{\lambda}{s + \lambda} \right)^{n-k}, \quad i \in S_0. \tag{5.72}$$

Using

$$\int_0^y x^m e^{ax} dx = e^{ay} \sum_{l=0}^{k} \frac{(-1)^l k! y^{k-l}}{(k-l)! a^{l+1}} - \frac{(-1)^k k!}{a^{k+1}}, \tag{5.73}$$

we have

$$\beta_i(y, s; n) = \frac{1}{\lambda} \sum_{k=0}^{n} c_{ni}(k) \left\{ \left(\frac{\lambda}{s + \lambda} \right)^{n+1} \right.$$

$$- e^{-sy} \sum_{l=0}^{k} \frac{e^{-\lambda y}(\lambda y)^{k-l}}{(k-l)!} \left(\frac{\lambda}{s+\lambda}\right)^{n-k+l+1} \Bigg\} , i \in S_0. \quad (5.74)$$

Inverting with respect to s, we obtain

$$
\begin{aligned}
\beta_i(y, s; n) &= \sum_{k=0}^{n} c_{ni}(k) \Bigg\{ \frac{e^{-\lambda t}(\lambda t)^n}{n!} \\
&\quad - \left(\sum_{l=0}^{k} \frac{e^{-\lambda y}(\lambda y)^{k-l}}{(k-l)!} \frac{e^{-\lambda(t-y)}[\lambda(t-y)]^{n-k+l}}{(n-k+l)!} \right) U(t-y) \Bigg\} \\
&= \frac{e^{-\lambda t}(\lambda t)^n}{n!} \Bigg\{ \sum_{k=0}^{n} c_{ni}(k) \times (1- \\
&\quad \sum_{l=0}^{k} \binom{n}{l} \left(\frac{y}{t}\right)^l \left(1-\frac{y}{t}\right)^{n-l} U(t-y)) \Bigg\} , i \in S_0. \quad (5.75)
\end{aligned}
$$

Since $U(t-y) + U(y-t) = 1$ and $\sum_{l=0}^{n} \binom{n}{l} \left(\frac{y}{t}\right)^l \left(1-\frac{y}{t}\right)^{n-l} = 1$, we have for $i \in S_0$:

$$
\begin{aligned}
\beta_i(y, s; n) &= \frac{e^{-\lambda t}(\lambda t)^n}{n!} \Bigg\{ \left(\sum_{k=0}^{n} c_{ni}(k) \right) U(y-t) \\
&\quad + \sum_{k=0}^{n} c_{ni}(k) \sum_{l=k+1}^{n} \binom{n}{l} \left(\frac{y}{t}\right)^l \left(1-\frac{y}{t}\right)^{n-l} U(t-y) \Bigg\} \\
&= \frac{e^{-\lambda t}(\lambda t)^n}{n!} \Bigg\{ \left(\sum_{k=0}^{n} c_{ni}(k) \right) U(y-t) \\
&\quad + \sum_{l=1}^{n} \binom{n}{l} \left(\frac{y}{t}\right)^l \left(1-\frac{y}{t}\right)^{n-l} \left(\sum_{k=0}^{l-1} c_{ni}(k) \right) \Bigg\}. \quad (5.76)
\end{aligned}
$$

Let

$$ d_{ni}(l) = \sum_{k=0}^{l} c_{ni}(k). \quad (5.77) $$

Then, we find the following expression for $\beta_i(y, s; n)$, for $i \in S_0$:

$$
\begin{aligned}
\frac{e^{-\lambda t}(\lambda t)^n}{n!} \Bigg\{ d_{ni}(n) U(y-t) + \\
\left[\sum_{l=1}^{n} \binom{n}{l} \left(\frac{y}{t}\right)^l \left(1-\frac{y}{t}\right)^{n-l} d_{ni}(l-1) \right] U(t-y) \Bigg\}. \quad (5.78)
\end{aligned}
$$

Proceeding in a similar manner, we find in case $i \in S_f$:

$$\beta_i(y, s; n) = \frac{e^{-\lambda t}(\lambda t)^n}{n!} \{d_{ni}(n)U(y - t)$$

$$+ \left[\sum_{l=0}^{n}\binom{n}{l}\left(\frac{y}{t}\right)^l\left(1 - \frac{y}{t}\right)^{n-l}d_{ni}(l)\right]U(t - y)\right\}. \quad (5.79)$$

The coefficients $c_{ni}(k)$ can be obtained from (5.49) as:

$$\begin{aligned}
c_{ni}(n) &= \sum_{j \in S_0} P_{ij}c_{n-1,j}(n - 1), \\
c_{ni}(0) &= \sum_{j \in S_f} P_{ij}c_{n-1,j}(0), \\
c_{ni}(k) &= \sum_{j \in S_0} P_{ij}c_{n-1,j}(k - 1) + \sum_{j \in S_f} P_{ij}c_{n-1,j}(k), \text{ for } k \neq 0, n.
\end{aligned}$$
$$(5.80)$$

The recursions are initiated with $c_{0i}(0) = 1$, for all $i = 1, 2, \cdots, N$. Using (5.77), the recursions for $d_{ni}(k)$ are:

$$\begin{aligned}
d_{ni}(0) &= \sum_{j \in S_f} P_{ij}d_{n-1,j}(0), \\
d_{ni}(k) &= \sum_{j \in S_0} P_{ij}d_{n-1,j}(k - 1) + \sum_{j \in S_f} P_{ij}d_{n-1,j}(k); \ k \neq n, 0, \\
d_{ni}(n) &= \sum_{j \in S_0} P_{ij}d_{n-1,j}(n - 1) + \sum_{j \in S_f} P_{ij}d_{n-1,j}(n - 1) = 1.
\end{aligned}$$
$$(5.81)$$

Using (5.81) in (5.78–5.79) and noting that $\beta_i(y, t) = \sum_{n=0}^{\infty} \beta_i(y, t; n)$, we obtain the conditional distribution of cumulative operational time:

$$\beta_i(\mu, s; n) = \begin{cases}
U(y - t) + \sum_{n=0}^{\infty} \frac{e^{-\lambda t}(\lambda t)^n}{n!} \\
\left[\sum_{k=1}^{n}\binom{n}{k}\left(\frac{y}{t}\right)^k\left(1 - \frac{y}{t}\right)^{n-k}d_{ni}(k - 1)\right]U(t - y), \ i \in S_0, \\
\\
U(y - t) + \sum_{n=0}^{\infty} \frac{e^{-\lambda t}(\lambda t)^n}{n!} \\
\left[\sum_{k=0}^{n}\binom{n}{k}\left(\frac{y}{t}\right)^k\left(1 - \frac{y}{t}\right)^{n-k}d_{ni}(k)\right]U(t - y), \ i \in S_f.
\end{cases}$$
$$(5.82)$$

An explicit expression for the kth moment is obtained as:

$$m_{ki}(t) = \int_0^{\infty} ky^{k-1}(1 - \beta_i(y, t))dy \quad (5.83)$$

$$= t^k\left[1 - \sum_{n=0}^{\infty} \frac{e^{-\lambda t}(\lambda t)^n}{n!}\left(\sum_{l=0}^{n} \frac{\binom{k+l-1}{l}}{\binom{n+k}{k}}d_{ni}(l)\right)\right]. \quad (5.84)$$

Chapter 6

Monotonicity and Error Bound Results

Nico M. van Dijk

A performability model, for instance of a database system subject to breakdowns or a communication network subject to link failures, can often be evaluated by means of Markov reward models. In this chapter it will be shown how such models can be used for three special purposes: (i) to establish transient monotonicity results; (ii) to compare a system under different disciplines, and (iii) to obtain a priori error bounds for approximate models.

The different purposes and possible results are illustrated for a simple representative front-end database system, a queueing network subject to breakdowns, and a circuit-switched system with link failures. These applications support the practical usefulness of the approach and suggest further exploitation.

6.1 Introduction

PERFORMABILITY evaluation deals with the combined aspects of *performance* and *reliability* (or availability) evaluation for computer and communication systems. Typical examples of such systems are computer networks in which one or more components such as a front end, a database, a memory, a switch or a processor can break down, and a communication network such as a circuit-switched network for combined voice and data transmissions in which communication links are subject to link failures.

As the periods during which such components or links are operable (up) usually depend on the state of neighbouring components and are usually subject to variability, the duration of these periods is normally assumed to be *random* according to some underlying distribution. A random or stochastic analysis is thus required in order to evaluate the performability of the system.

Continuous-time Markov chains provide a general stochastic evaluation tool based on the (approximative) assumption of an underlying exponential structure. Though often not satisfied in reality, the exponentiality assumption appears justifiable for approximation purposes as it enables one to reduce computations to standard matrix computational problems (e.g., see Trivedi *et al.* [465]). In addition, on some occasions it may even lead to closed-form analytic expressions that are valid also for non-exponential cases (e.g., see Smeitink *et al.* [438]).

However, as the computational complexity of these Markov chains can still be prohibitively large while the necessary conditions for these closed-form solutions are often not satisfied, a number of questions still remain in order to evaluate a performability model either by an exact numerical computation or by a simplified approximate model. Three such major questions are:

1. How can we know and prove whether a performance function increases (or decreases) monotonically in time, as time-dependent results can hardly be obtained?

2. When comparing a system under different disciplines, most notably to simplify the computations or to justify a simple analytic expression, how can we compare the performance under these different disciplines?

3. In situations as under point 2, or when ignoring some complicating aspects, how can we provide secure (analytic *a priori*) error bounds on the inaccuracy introduced by the simplification?

This chapter aims to show and illustrate how these three general questions can be addressed more or less in a unified approach by exploiting discrete-time Markov chain properties combined with one-step Markov reward or dynamic programming arguments.

Three general results are therefore presented. These results, in fact, apply to a wider setting and have been obtained by combination of recently published results for arbitrary Markov chains (*cf.* [121, 122, 132]). However, their general application to performability analysis has not yet been reported and shown to be possible.

Furthermore, some special performability modelling applications are studied in detail which involve technicalities that have not been dealt with before, most notably:

- *Monotonicity* results are shown for systems with *dependent components*.

- *An error bound* is established for *ignoring more than one breakdown*.

Further extensions such as to non-exponential systems and other applications are hereby supported.

Comparison with literature The topic of stochastic monotonicity has received considerable attention over the last decade, motivated by the pioneering work of Stoyan [446]. Some notable references here are: Keilson and Kester [258], Massey [302] and Whitt [476, 477]. These results are also directly related to sample path results in combination with weak coupling arguments. Typical applications are comparison results for stochastic service or queueing networks as have been extensively studied over the last decade (Adan and van der Wal [2, 3], Shanthikumar and Yao [432, 434, 433], Tsoucas and Walrand [466], van Dijk, Tsoucas and Walrand [129]).

So far, however, only a few results using this technique have been reported for performability models. Two elegant exceptions are the papers by Keilson and Kester [198] and Massey [302] which report monotonicity results for systems with components that can only break down *independently*. In these two references general conditions are provided in terms of transition rates for a system to be stochastically monotone. For a concrete performability modelling system with dependent components, however, the verification of these conditions is still open and can be rather complicated. Furthermore, the conditions imply strong stochastic monotonicity results, that is on a sample path basis, whereas performance measures of interest such as a throughput or point availability only concern an arbitrary marginal instant.

As a first step, an approach is therefore suggested that requires monotonicity to be proved only for *specific measures of interest at marginal instants.*

In other situations one might wish to consider a system under slightly modified (simplified) circumstances so as to obtain a simple expression. To this end, in van Dijk [121] a general error bound theorem was developed as an extension of an idea developed in van Dijk and Puterman [132]. The essence of this theorem is to analyze steady state measures by cumulative Markov reward structures and to apply inductive Markov reward arguments to estimate so-called bias terms. This technique has already been applied successfully to a number of queueing network situations also without a product form (*cf.* [118, 119, 131, 130]). Theorems 6.2 and 6.3, presented later in this chapter, are similar in nature except that a direct weighting of the steady state probabilities in Theorem 6.3 is included.

Outline First, in Section 6.2, preliminary results are given in order to deal with continuous-time Markov reward models. Furthermore, a motivation performability example is outlined. Then, in Section 6.3, the question of the time-monotonic behaviour of performability measures is addressed. A general monotonicity result is established and illustrated by an application. In Section 6.4, the Markov reward approach is applied to obtain comparison results. The essential differences from stochastic comparison approaches are discussed as well as made concrete by the technical steps for a performability application. In Section 6.5 the major usage of the Markov reward approach is described, that is, how it enables one to conclude error bounds for system modifications. It is then applied to two performability models. Section 6.6 summarizes the chapter.

6.2 Preliminaries and a motivating example

In this section we first provide some preliminary results as a common basis and background for the successive sections. This particularly involves the uniformization step to deal with the continuous-time models in a discrete-time manner. An example is described in detail to motivate our study.

6.2.1 Marginal expectations

Throughout we will consider continuous-time Markov chains (CTMC) with countable state space S and generator matrix $\mathbf{Q} = \mathbf{q}(i, j)$, the transition rate for a change from state i into state j. For convenience this chain is assumed to be *uniformizable*. That is, for some finite constant Λ and for all $i \in S$:

$$\sum_{j \neq i} \mathbf{q}(i, j) \leq \Lambda. \tag{6.1}$$

Let $\mathbf{P}_t(i, j)$ denote the transition probability from state i into state j over time t and define expectation operators $\{\mathbf{T}_t \mid t \geq 0\}$ on the set B of real-valued functions f defined on S by:

$$(\mathbf{T}_t f)(i) = \sum_j \mathbf{P}_t(i, j) f(j). \tag{6.2}$$

In words, $(\mathbf{T}f)(i)$ represents the expected value of function f at time t of the CTMC when starting in state i at time 0. By virtue of the boundedness (uniformization) assumption (6.1) it is then well known (e.g., [194, 198]) that the continuous-time Markov chain can also be evaluated as a discrete-time Markov chain (DTMC) with one-step transition matrix:

$$\mathbf{P} = \mathbf{I} + \mathbf{Q}/\Lambda,$$

or more precisely, with one-step transition probalities

$$\mathbf{P}(i, j) = \begin{cases} \mathbf{q}(i, j)/\Lambda, & j \neq i, \\ 1 - \sum_{j \neq i} \mathbf{q}(i, j)/\Lambda, & j = i. \end{cases} \tag{6.3}$$

Intuitively, one may regard this matrix as a transition matrix over a time interval of length $\Delta = 1/\Lambda$. In contrast with the CTMC, however, this ignores possible multiple changes in this time interval. Nevertheless, it can be shown that the stochastic behaviour of the CTMC, or more precisely the transition mechanisms and corresponding expectation over any time t, can stochastically be obtained as if at exponential times with parameter Λ; hence on average per time interval of length $\Delta = 1/\Lambda$, a change may take place according to the one-step transition matrix $\mathbf{P}$. This is expressed by the following relation, where $\mathbf{T}^k$ for the DTMC represents (similarly to $\mathbf{P}_t$ for the CTMC) the expectation operator over k steps, $\mathbf{P}^k$ denotes the kth matrix power of $\mathbf{P}$, and $\mathbf{I}$ is the identity operator:

$$\begin{cases} \mathbf{T}_t f(i) & = \sum_{k=0}^{\infty} \frac{(t\Lambda)^k}{k!} e^{-t\Lambda} \quad \mathbf{T}^k f(i), \quad i \in S, \\ \mathbf{T}^k f(i) & = \sum_j \mathbf{P}^k(i,j) f(j); \quad \mathbf{T}^0 = \mathbf{I}, \quad \text{for all } f \in B. \end{cases} \tag{6.4}$$

As a major advantage of this uniformization we are now able to study properties for $\mathbf{T}_t$ by studying related properties for $\mathbf{T}^k$. A first point of attention in this direction will be monotonicity properties of the expectation operators $\mathbf{T}_t$ as a function of t. Particularly for performability applications, here one may think of monotone convergence to a steady state performance measure, $\mathbf{G}$, as by

$$\mathbf{T}_t r(i) \to \mathbf{G}, \quad \text{as } t \to \infty, \tag{6.5}$$

where $\mathbf{G}$ represents the expected performability in the long run for some given performance rate function $r(.)$, such as measuring the number of components that are up and the initial state i, such as a starting state of the system in perfect condition.

6.2.2   Cumulative measures

To compare two different CTMCs, for instance where one of them might be a simplified modification of the other, it will be convenient to use cumulative performance measures. This also applies when the actual interest concerns a marginal performance measure, most notably a measure associated with steady state probabilities.

To this end, consider some given reward rate function $r(i)$ that incurs a reward $r(i)$ per unit of time whenever the system is in state i. The expected **cumulative** reward over a period of length t and given the initial state i at time 0 is then given by:

$$\mathbf{V}_t(i) = \int_0^t \mathbf{T}_s r(i) ds. \tag{6.6}$$

Then, as in (6.5), under natural ergodicity conditions this cumulative measure averaged over time will convergence to some average reward, or in the current setting, some performability measure $\mathbf{G}$, independently of the initial state i, as:

$$\frac{1}{t}\, \mathbf{V}_t(i) \to \mathbf{G}, \quad t \to \infty. \tag{6.7}$$

Now by virtue of the uniformization technique, we can also evaluate $\mathbf{G}$ by means of expected cumulative rewards for the uniformized discrete-time Markov chain as:

$$\frac{\Lambda}{k}\, \mathbf{V}^k(i) \to \mathbf{G}, \quad k \to \infty, \tag{6.8}$$

where $\mathbf{V}^k(i)$ represents the expected cumulative reward for the uniformized DTMC over k steps, each of mean length $1/\Lambda$, with one-step rewards $r(j)/\Lambda$ per step whenever the system is in state j:

$$\mathbf{V}^k = \frac{1}{\Lambda} \sum_{s=0}^{k-1} \mathbf{T}^s, r \qquad \mathbf{V}^0 = r. \tag{6.9}$$

Here one may state that the factor Λ in (6.8) is required as the time average of $\mathbf{V}^k$ ensures an average reward per step of mean length $1/k$ instead of per unit of time.

The major advantage of this discrete setup is that it enables one to use inductive arguments by exploiting the reward (or dynamic programming) relation:

$$\mathbf{V}^{k+1}(i) = \frac{r(i)}{\Lambda} + \sum_j \mathbf{P}(i,j)\, \mathbf{V}^k(j), \qquad k = 0, 1, 2, \cdots, \quad i \in S. \tag{6.10}$$

In words, the expected cumulative reward over $k+1$ steps can be obtained by first considering the immediate one-step reward incurred in the first step and next by adding the expected cumulative reward over the remaining k steps onward after having made one transition.

6.2.3 A performability example

To motivate and illustrate our interest, in this section we present a simple performability example of a *front-end database system* (Figure 6.1). This system consists of a front-end (*FE*), a database (*DB*), and two subsystems (*SS*$_1$ and *SS*$_2$) each in turn containing one switch (*S*$_i$), one memory (*M*$_i$) and two processors (*P*$_{i1}$ and *P*$_{i2}$).

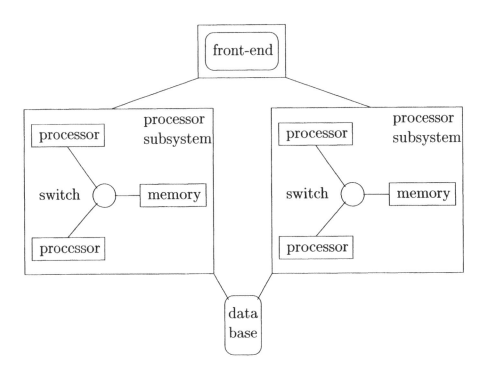

Figure 6.1: A front-end database system

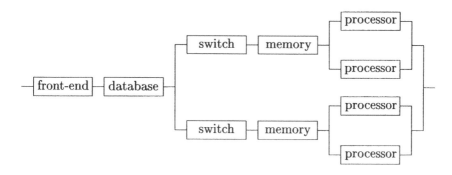

Figure 6.2: Reliability block diagram for the database system

Each of the individual components is subject to breakdowns which render it inoperative. The whole system is said to be operable (up) only under the following conditions:

- both the *FE* and *DB* are operable (up);

- at least one of the two subsystems is operable (up).

Similarly, a subsystem i is said to be operable (up) only under the following conditions:

- both the switch S_i and memory M_i are operable (up);

- at least one of the two processors P_{i1} or P_{i2} is operable (up).

In other words, the operability of the system is hierarchical as visualized by the reliability block diagram (see Figure 6.2) with the components ordered with regard to their importance for the system to be operable. In all other cases the system is said to be out of operation, during which time none of the components can break down as they are not working. Assume that the components are numbered $1, \cdots, 10$ (in some arbitrary way) and that each component $h \in \{1, 2, \cdots, 10\}$ can independently break down at rate β_h. Furthermore, assume that there is just one repair facility.

The way the repair facility operates will determine the actual system behaviour. For example, the existence of a simple closed-form expression for the steady-state distribution of down components depends on whether or not repair priorities are assigned, as will be explained in the following subsection. So, the question is: *do we have a product-form or not?* To answer this question, we distinguish two cases.

Case 1. Repair in order of breakdown When the repair facility is fully assigned to the components in order of their breakdowns, observe that one might get into a state in which both component 1 (FE) and component 3 (switch SS_1) are down while the repair facility resides at this last component 3. In state (1,3), where a state represents the components that are down, we have:

$$\text{Rate into state } (1,3) \text{ due to component 1 by } (3) \Rightarrow (1,3) > 0$$
$$\text{Rate out of state } (1,3) \text{ due to component } 1 \text{ by } (1,3) \Rightarrow (3) = 0$$
$$(6.11)$$

A notion of (flow) *balance per individual component* thus necessarily fails in state (1,3). As shown in [386], such a property must hold in order to conclude a so-called product-form expression. Indeed, for the present system *a simple analytic expression* does not exist.

Case 2. Repair priority in hierarchical order In light of the balance inconsistency (6.11) for the FE, and similarly the DB, the following modification could be suggested to resolve the inconsistency:

> *At system level the* **repair** *should be assigned with* **priority** *for the* **FE** *or* **DB** *at the very instant that one of these breaks down. Repair that might take place at* $\qquad$ (6.12) *one of the subsystems should hereby be interrupted to be resumed later on.*

By similar reasoning, also within a subsystem priorities should be assigned:

> *At subsystem level the* **repair** *facility should be assigned with* **priority** *for* **S** *or* **M** *at the very instant that one of these breaks down. Repair that might take place at* $\qquad$ (6.13) *one of the processors should hereby be interrupted to be resumed later on.*

> *Furthermore, when both subsystems are down, assume that each receives half the amount of the repair capacity and, similarly, when both processors in a subsystem are* $\qquad$ (6.14) *down each receives half the amount of the repair capacity of that subsystem.*

Balance inconsistencies at component level as in (6.11) then seem to be repaired which, according to [386], would guarantee a product-form expression. Indeed, under (6.12), (6.13) and (6.14) one can verify the following product-form expression by verifying balance equations per individual component, where the state $H = h_1, \cdots, h_n$ represents the components that are down, c is a normalizing constant, β_h is the exponential breakdown parameter and δ_h the exponential repair time parameter of component h:

$$\pi(H) = cf(H) \prod_{i \in H} \left[\frac{\beta_i}{\delta_i} \right], \qquad (6.15)$$

where $f(H)$ is a repair scaling factor which can be shown to be given by:

$$f(H) = \begin{cases} 1, & \text{if } \textit{at most one} \text{ subsystem is down,} \\ & \text{whereas, if } \textit{both} \text{ subsystems are down:} \\ 2, & \text{if in } \textit{neither} \text{ of these two, both processors are down,} \\ 4, & \text{if in } \textit{one} \text{ of these two, both processors are down,} \\ 8, & \text{if in } \textit{both} \text{ of these two, both processors are down.} \end{cases}$$

$$(6.16)$$

Modification Whether Case 1 or Case 2 above is in order in a practical situation depends on the actual application. Case 1, however, may suggest an artificial modification in Case 2 in order to facilitate computation. This modification might still provide a reasonable first indication of orders of magnitude as well as secure performance bounds.

Questions Various questions of interest are hereby generated:

1. **Monotonicity.** How useful is a steady-state analysis as one is often interested in performance at specific instants? We aim to provide conditions to guarantee that steady-state measures can also be useful in these situations as secure bounds. To this end, we will develop monotonicity results.

2. **Comparison.** Does the modification approach indeed lead to secure performance bounds? In other words, when comparing two systems or a system under two different operations, can we formally prove that one performs better than the other with respect to some measure?

3. **Error bounds.** Alternatively, in situations as under 2, can we provide an analytic error bound on the accuracy of a simplified approximative expression?

In what follows, we will address these three questions in a somewhat more abstract setting of Markov chains, but tailored to performability applications.

6.3   Monotonicity in time

In this section we develop conditions and illustrate their application to a performability setting by which one can conclude whether a performability is monotonically decreasing or increasing in time. Such results are directly related to stochastic monotonicity results as have been intensively studied over the last decades, most notably by Stoyan [446], Whitt [476, 477], Keilson and Kerster [258] and Massey [302].

These references, however, all provide conditions in terms of monotonicity of the generator or transition rates of the underlying Markov chain, or, relatedly, conditions that the underlying process is monotone with respect to sample path properties. However, other than for simple performability models with *independent* components (see Keilson and Kester [258] and Massey [302]), the verification of these conditions has not been touched upon for more realistic performability models, such as with dependent breakdowns and repairs. Therefore, this is the focus of the present section.

The analysis in this section therefore is kept self-contained and in essence is mainly concerned with the actual verification of the necessary monotonicity conditions in a concrete performability situation with dependent components. Such results have not yet been reported. Particularly, Theorem 6.1 is therefore of interest.

In Section 6.3.1, the performability model of interest in this section will be described. Then, in Section 6.3.2, a general time-monotonicity result for this model is obtained (Theorem 6.1), which involves the verification of a technical condition (Lemma 6.1). An example is given in Section 6.3.3.

6.3.1   A performability model

Consider a performability model which consists of N components numbered $1, \cdots, N$ each of which can alternatively be in an *up* and a *down* state. More precisely, let the state $H = \{h_1, h_2, \cdots, h_n\}$ represent components $h_1, h_2, \cdots, h_n$ being down, and for $h \in H$ let $H - h$ denote the same state with h deleted and for $h \notin H$ let $H + h$ denote the same state with h added. Then we define the functions:

$$\begin{cases} \beta(h|H), & \text{breakdown rate} \quad h \notin H, \\ \delta(h|H), & \text{repair rate} \quad\quad\; h \in H, \end{cases} \tag{6.17}$$

with the interpretation that an up component $h \notin H$ when the system is in state H will break down, by which the state will change into $H + h$ at a rate $\beta(h|H)$, while a down component $h \in H$ when the system is in state H will be repaired at a rate $\delta(h|H)$. Here both $\beta(.|.)$ and $\delta(.|.)$ can be zero. Furthermore, let $r(H)$ be a performance rate or reward in state H. For example, by $r(H) = |H| = n$, we measure the total number of down components, as a measure of performance. Adapting the general notation from Section 6.2, where we identify a state H as a state i, we are interested in the expected performance, e.g., the performability, at an arbitrary instant t given by:

$$\mathbf{A}_t = \sum_H P(X_0 = H) \, \mathbf{T}_t r(H), \tag{6.18}$$

where X_0 is some given initial distribution at time $t = 0$. In particular, we are interested in the possible monotonic behaviour of A_t as a function of t, for appropriate initial distribution X_0.

6.3.2   Time monotonicity results

Let $r(.)$ be an arbitrary function such that

$$r(H + h) \geq r(H), \quad\quad \text{for all } H \text{ and } h \notin H, \tag{6.19}$$

so that $r(.)$ can be seen as a measure of performability. Furthermore, we make the natural assumptions that:

$$\begin{cases} \beta(s|H + h) \geq \beta(s|H), & s \notin H + h, \\ \delta(s|H + h) \leq \delta(s|H), & s \in H. \end{cases} \tag{6.20}$$

In words, the likelihood of another breakdown of an individual component may become larger the more components are down, and the repair speed for an individual component may become smaller the more components are in repair. Under these assumptions we can prove the following intuitively appealing result (see Section 6.3.3 for a specific application). Essential for this result to be proven is that the one-step transition mechanism or rather the underlying transition structure of the CTMC preserves monotonicity properties, as will be shown in Lemma 6.1 afterwards.

Theorem 6.1 *When starting with a perfect system, that is $P(X_0 = \emptyset) = 1$, and under the conditions (6.19) and (6.20) we have:*

$$\mathbf{A}_t \downarrow \mathbf{A}_\infty. \tag{6.21}$$

Proof. First we will prove that

$$\mathbf{T}^k f(\emptyset) \geq \mathbf{T}^{k+1} f(\emptyset), \qquad k \geq 0, \tag{6.22}$$

for arbitrary monotone functions $f \in M$ where

$$M = \{f : S \to \mathbb{R} | f(H + h) \geq f(H) \text{ for all } H + h \in S\}. \tag{6.23}$$

To this end, for $f \in M$ we can write:

$$\mathbf{T}^{k+1} f(\emptyset) - \mathbf{T}^k f(\emptyset) = \mathbf{T}^k (\mathbf{T}f)(\emptyset) - \mathbf{T}^{k-1}(\mathbf{T}f)(\emptyset). \tag{6.24}$$

By virtue of Lemma 6.1 below, the proof of (6.22) would thus be completed by induction to k, provided

(i) (6.22) holds for $k = 0$;

(ii) $\mathbf{T}f \in M$ for $f \in M$.

To prove (i) and (ii), first note that the uniformization matrix $\mathbf{P}$ as per (6.3) must have been chosen with M such that for all $H \in S$:

$$\Lambda \geq \sum_{s \notin H} \beta(s|H) + \sum_{s \in H} \delta(s|H). \tag{6.25}$$

The proof of (i) now follows directly by substituting the uniformization matrix (6.3), using (6.25) and noting that $f(s) \geq f(\emptyset)$ for $f \in M$, and writing:

$$\mathbf{T}f(\emptyset) = \Lambda^{-1} \sum_s \beta(s|\Phi)f(s) + [1 - \Lambda^{-1} \sum_s \beta(s|\Phi)]f(0) \geq f(\Phi). \tag{6.26}$$

The *proof of (ii)* is given in Lemma 6.1 below. The proof of (6.22) is hereby completed for any $f \in M$, in particular for $f = r$.

To complete the proof of (6.21), now recall expression (6.4) with $i = \emptyset$ and compare the values $\mathbf{T}_t r(\emptyset)$ and $\mathbf{T}_{t+u} r(\emptyset)$ for arbitrary $t, u > 0$. Let $\alpha = tM$ and $\beta = (t+u)M$ be the corresponding Poisson parameters arising in their Poissonian expansions (6.4). As $\alpha < \beta$, we directly conclude by standard calculus (or probabilistic interpretation):

$$\sum_{k=0}^{m} \frac{\alpha^k}{k!} e^{-\alpha} \geq \sum_{k=0}^{m} \frac{\beta^k}{k!} e^{-\beta}, \qquad \text{for some } m > 0. \tag{6.27}$$

In combination with the fact that

$$\sum_{k=0}^{\infty} \frac{\alpha^k}{k!} e^{-\alpha} = \sum_{k=0}^{\infty} \frac{\beta^k}{k!} e^{-\beta} = 1, \tag{6.28}$$

so that any difference between the left- and right-hand sides in (6.27) will have to be compensated by higher k values, and also that, as per (6.22):

$$\mathbf{T}^k r(\emptyset) \leq \mathbf{T}^{k+1} r(\emptyset), \tag{6.29}$$

we can conclude from the Poisson expansion (6.4) and the discrete monotonicity (6.22):

$$\mathbf{T}_t r(\emptyset) \leq \mathbf{T}_{t+u} r(\emptyset). \tag{6.30}$$

With $\mathbf{A}_t$ the expected availability at time t as per Section 6.3.1, this completes the proof. □

Lemma 6.1 *The set of monotone functions* M *is closed under* T*. That is,*

$$\mathbf{T} f \in M \text{ for any } f \in M. \tag{6.31}$$

Proof. By substituting the uniformization matrix as per (6.3) we obtain:

$$
\begin{aligned}
\mathbf{T} f(H + h) \;=\; & \frac{1}{\Lambda} \sum_{s \notin H+h} \beta(s|H+h) f(H+h+s) \\
& + \frac{1}{\Lambda} \sum_{s \in H+h} \delta(s|H+h) f(H+h-s) \\
& + \left[1 - \frac{1}{\Lambda} \sum_{s \notin H+h} \beta(s|H+h) - \frac{1}{\Lambda} \sum_{s \in H+h} \delta(s|H+h) \right] f(H+h)
\end{aligned} \tag{6.32}
$$

and similarly:

$$\mathbf{T}f(H) = \Lambda^{-1}\sum_{s\notin H}\beta(s|H)f(H+s) + \Lambda^{-1}\sum_{s\in H}\delta(s|H)f(H-s)$$

$$+ \left[1 - \Lambda^{-1}\sum_{s\notin H}\beta(s|H) - \Lambda^{-1}\sum_{s\in H}\delta(s|H)\right]f(H). \qquad (6.33)$$

In expressions (6.32) and (6.33) now substitute:

$$\begin{cases} \beta(s|H+h) = \beta(s|H) + [\beta(s|H+h) - \beta(s|H)], \\ \delta(s|H) = \delta(s|H+h) + \delta(s|H) - \delta(s|H+h)]. \end{cases} \qquad (6.34)$$

In addition, artificially *add* and *substract* the following terms:

$$\Lambda^{-1}\beta(h|H)f(H+h) + \Lambda^{-1}\sum_{s\in H}[\delta(s|H) - \delta(s|H+h)]f(H+h)$$

in (6.32) and

$$\Lambda^{-1}\delta(h|H+h)f(H) + \Lambda^{-1}\sum_{s\notin H+h}[\beta(s|H+h) - \beta(s|H)]f(H)$$

in (6.33). By making these substitutions and by subtracting (6.33) from (6.32), we obtain:

$$\mathbf{T}f(H+h) - \mathbf{T}f(H) =$$
$$\Lambda^{-1}\sum_{s\notin H+h}\beta(s|H)[f(H+h+s) - f(H+s)] +$$
$$\Lambda^{-1}\sum_{s\notin H+h}\beta(s|H) - \beta(s|H)][f(H+h+s) - f(H)] +$$
$$\Lambda^{-1}\beta(h|H)[f(H+h) - f(H+h)] +$$
$$\Lambda^{-1}\sum_{s\in H}\delta(s|H+h)[f(H+h-s) - f(H-s)] +$$
$$\Lambda^{-1}\sum_{s\in H}\delta(s|H) - \delta(s|H+h)][f(H+h-s) - f(H-s)] +$$
$$\Lambda^{-1}\delta(h|H+h)\{f(H) - f(H)\} +$$
$$\left(1 - \frac{1}{\Lambda}\left\{\sum_{s\notin H+h}\beta(s|H+h) - \beta(h|H) - \sum_{s\in H}\{\delta(s|H) - \delta(h|H+h)\}\right\}\right)$$
$$\times[f(H+h) - f(H)].$$

As $f(H+h) = [f(H+h)-f(H)]+[f(H)-f(H-s)]+f(H-s)$, application of (6.23) thus implies that $\mathbf{T}f(H + h) - \mathbf{T}f(H) \geq 0$, which completes the proof. □

6.3.3   A performability application

As transient analysis is usually (at least) computationally expensive to perform, Theorem 6.1 can be useful to justify an analysis on a steady-state basis instead. Particularly, when *closed-form* expressions such as those of the form (6.15) or more general related forms as reported in [438] can be concluded in the steady state case, either for the system under consideration itself (Case 1 in Section 6.2.3) or for a slightly modified system (Case 2 in Section 6.2.3), Theorem 6.1 seems appealing to guarantee a secure **lower** or **upper bound** for a transient performability measure $\mathbf{A}_t$.

These bounds furthermore can be quite reasonable taking into account the interpretation of up and down components and the fact that fractions of the time during which components are down should realistically be small. Let us give one example.

Reconsider the performability example from Section 6.2.3 under Case 2 and recall that H denotes the set of components that are down. Then in the present parametrization:

$$\begin{cases} \beta(h|H) = \beta_h, & \text{independent breakdown,} \\ \delta(s|H + h) \leq \delta(s|H), & \text{dependent repair,} \end{cases} \tag{6.35}$$

for all admissible states $H + h$, $H \in S$. Further let $r(H) = [|H|/N] = [n/N]$ to measure the fraction of down components. The conditions (6.19) and (6.20) are hereby satisfied. By virtue of Theorem 6.1, the expected number of down components when starting with a perfect system (state $\emptyset$) thus converges, monotonically increasing to the steady-state value $\mathbf{A}_\infty$ as illustrated in Figure 6.3. This steady-state value $\mathbf{A}_\infty$ in turn can directly be calculated as by:

$$\mathbf{A}_\infty = \sum_H \left[\frac{n}{N}\right] \pi(H), \qquad \text{with } \pi(H) \text{ as given in (6.15).} \tag{6.36}$$

For example, when $\mathbf{A}_\infty = 96\%$, a secure lower bound of 96% is provided for the transient availability function at $\mathbf{A}_t$ any instant t.

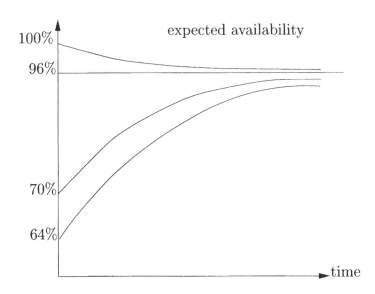

Figure 6.3: Monotonic transient availability curves

6.4   Comparison results

In this section the Markov reward approach is used to establish comparison results. In Section 6.4.1 the advantages but also the disadvantages of this approach over the standard stochastic comparison approach will be listed briefly. In Section 6.4.2 the general comparison theorem is given. The verification of its technical conditions as well as the essence of this approach over the stochastic comparison approach on some occasions is illustrated in detail in Section 6.4.3, using queueing networks subject to breakdowns.

6.4.1   Introduction

We provide two general theorems when aiming to compare a cumulative or steady state performance measure between two systems, where typically one system might be a *perturbation, modification* or *truncation* of the other or where both systems slightly differ in conditions such as starting states. These theorems are a combination of two results adopted from [121] and [122] and are based on *Markov reward theory*. Over the last decades, comparison results have been intensively studied by means of *sample path* and weak coupling results (e.g., [2, 3, 432, 434, 433, 466, 460]). For comparison purposes (for another purpose see Section 6.5), the Markov reward approach has advantages and disadvantages over the sample path approach. As no such comparison between them has been reported, let us briefly summarize these (dis)advantages without stepping into details. Advantages of the sample path approach are:

- *Results* obtained with it *are stronger* since it (i) applies just as well *without exponentiality assumptions*, and (ii) provides statements on a *with probability 1 sample path* basis.

Advantages of the Markov reward approach are:

- It provides statements just for *expected measures*. In particular, it may apply just for *marginal (instant) expectations*.

- As a consequence, the necessary underlying system *conditions* can be *weaker*. In particular, the Markov reward approach may apply where the sample path approach will *not work*. Natural examples can be given where the necessary stochastic monotonicity or equivalent sample path conditions necessarily fail, but where monotonicity or comparison for the specific measure can still be proven (e.g., [128, 130]). Also see the application in Section 6.4.3.

- Even more, comparison results on a *steady state basis* can be obtained where comparison or monotonicity results on a sample path basis (in the strong or the weak sense) fail (see Section 6.4.3).

- It can be tailored to just one (or a category of) *specific performance measure*.

6.4.2 General comparison lemma

Consider a CTMC as described in Section 6.2 with transition rates $\mathbf{q}(i,j)$, reward rates $r(i)$ and state space S. We briefly denote this parametrization by $(S, \mathbf{q}, r)$.

Now consider a second CTMC, described similarly, that can be thought of as a modified version of the first, $(\bar{S}, \bar{\mathbf{q}}, \bar{r})$. In short, we aim to compare these two models:

$$\begin{cases} (S, \mathbf{q}, r) \\ (\bar{S}, \bar{\mathbf{q}}, \bar{r}) \end{cases} \quad \text{under condition } \bar{S} \subset S. \tag{6.37}$$

Throughout, we use the overbar symbol for an expression concerning the second CTMC and the overbar in parentheses to indicate that the expression is to be read for both CTMCs. We aim to compare the performance of the two CTMCs in terms of their expected cumulative reward functions $\mathbf{V}_t$ and their expected reward per unit time in steady state $(\mathbf{G})$. To this end we provide the following theorem. In order for this chapter to be self-contained we copy a part of the proof from [446].

Theorem 6.2 *Suppose that for all $i \in \bar{S}$ and k:*

$$[\bar{r} - r](i) + \sum_j [\bar{\mathbf{q}}(i,j) - \mathbf{q}(i,j)][\mathbf{V}^k(j) - \mathbf{V}^k(i)] \geq 0, \tag{6.38}$$

then $\bar{\mathbf{V}}_t(l) \geq \mathbf{V}_t(l)$, for all t and $l \in \bar{S}$ and $\bar{\mathbf{G}} \geq \mathbf{G}$.

Proof. By virtue of (6.10) we have:

$$\begin{aligned}
\mathbf{V}^{k+1}(i) &= r(i)\Lambda^{-1} + \mathbf{T}\mathbf{V}^k(i), \\
\bar{\mathbf{V}}^{k+1}(i) &= \bar{r}(i)\Lambda^{-1} + \bar{\mathbf{T}}\bar{\mathbf{V}}^k(i).
\end{aligned} \tag{6.39}$$

As the transition probabilities $\bar{P}(.,.)$ remain restricted to $\bar{S} \subset S$, for arbitrary $l \in \bar{S}$ we can write the following for $\bar{\mathbf{V}}^k - \mathbf{V}^k(l)$:

$$\begin{aligned}
(\bar{r} - r)(l)\Lambda^{-1} + (\bar{\mathbf{T}}\bar{\mathbf{V}}^k - 1 - \mathbf{T}\mathbf{V}^{k-1})(l) &= \\
(\bar{r} - r)(l)\Lambda^{-1} + (\bar{\mathbf{T}} - \mathbf{T})\mathbf{V}^{k-1}(l) + \bar{\mathbf{T}}(\bar{\mathbf{V}}^{k-1} - \mathbf{V}^{k-1})(l) &= \quad (6.40) \\
\sum_{s=0}^{k-1} \left\{ \bar{\mathbf{T}}^s[\bar{r} - r](l)\Lambda^{-1} + \bar{\mathbf{T}}^s[(\bar{\mathbf{T}} - \mathbf{T})\mathbf{V}^{k-s-1}](l) \right\} + \bar{\mathbf{T}}^k(\bar{\mathbf{V}}^0 - \mathbf{V}^0)(l),
\end{aligned}$$

where the last step follows by iteration. First note that the last term in the latter right-hand side is equal to 0 as $\bar{V}^0(.) = V^0(.) = 0$. Further, by (6.3) and (6.4) again, we can also write for $(\bar{\mathbf{T}} - \mathbf{T})\mathbf{V}^s(i)$:

$$\begin{aligned}
\sum_{j \neq i}[\bar{\mathbf{q}}(i,j) - \mathbf{q}(i,j)]\Lambda^{-1}\mathbf{V}^s(j) - \sum_{j \neq i}[\bar{\mathbf{q}}(i,j) - \mathbf{q}(i,j)]\Lambda^{-1}\mathbf{V}^s(i) &= \\
\sum_{j \neq i}[\bar{\mathbf{q}}(i,j) - \mathbf{q}(i,j)][\mathbf{V}^s(j) - \mathbf{V}^s(i)]\Lambda^{-1}. \quad (6.41)
\end{aligned}$$

By substituting (6.41) and noting that $\bar{T}^s$ is a monotone operator for all s, i.e., $\bar{T}^s f \leq \bar{T}^s f$ if $f \leq g$ component-wise, we then obtain from (6.40) by using (6.38):

$$(\bar{\mathbf{V}}^k - \mathbf{V}^k)(l) = \sum_{s=0}^{k-1} \bar{\mathbf{T}}^s\{[\bar{r} - r] + (\bar{\mathbf{T}} - \mathbf{T})\mathbf{V}^{k-s-1}\}(l) \geq 0. \tag{6.42}$$

To conclude the proof, by standard calculus (see Lemma 2.1 in [122], (6.42) and (6.7)), we can write V_t as:

$$\mathbf{V}_t(l) = \sum_{k=1}^{\infty} e^{-t\Lambda} \frac{(t\Lambda)^k}{k!} \, \mathbf{V}^k(l). \qquad (6.43)$$

□

Remark 1: Essential differences from the stochastic comparison method Roughly speaking, with the stochastic comparison, or relatedly the sample path approach, as intensively studied in the literature (e.g. [258, 302, 446]), one essentially proves that the one-change transition structure or rather the transition rate matrices $\bar{\mathbf{Q}}$ and $\mathbf{Q}$ are directly ordered as $\bar{\mathbf{Q}} \geq (\leq)\mathbf{Q}$ in some appropriate ordering sense.

In the present setting this would mean that condition (6.38) is implied *without* the (one-step) *reward term* $[\bar{r} - r]$. However, this strong form of required ordering will not always be satisfied in practical applications, for which purpose the extra reward term $[\bar{r} - r]$ in condition (6.38) might give compensation. The application in the next subsection will contain this phenomenon which will be elaborated upon further below. In addition, by a simple sample path example, it will be illustrated that a sample path comparison may fail to prove the desired results on a *realization basis*. By the Markov reward approach as employed herein, however, this will be proven on an *expectation basis*.

Remark 2: Bias terms To apply Theorem 6.2 one needs to bound the so-called bias terms $\mathbf{V}^k(j) - \mathbf{V}^k(i)$ from below and/or above in the application under consideration. This bounding will be a technical complication but can generally be performed in an inductive manner by using the recursive relation (6.10). It has already successfully been applied to a number of complex queueing network situations for which no exact (e.g. product-form) expression could be found (*cf.* [119, 131, 130]). In the next subsection this will be illustrated for a particular application tailored to performability.

6.4.3   A special performability application

Consider a Jackson queueing network with N service stations (see Figure 6.4), Poisson arrivals with parameter λ, routing probalities p_{ij} from station i to station j, with p_{oj} for an arriving job to enter station j and $p_{io} = [1 - \sum_j p_{ij}]$ for a job to leave the system when leaving station i. The service rate at station i is $\mu_i(n_i)$ when n_i jobs are present, where $\mu_i(0)$ is assumed to be nondecreasing. Furthermore, the system has a capacity constraint for no more than F jobs in total. When F jobs are present, an arriving job is rejected and lost.

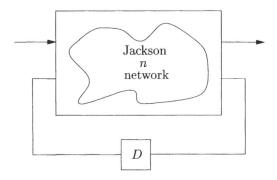

Figure 6.4: The considered Jackson queueing network

In addition, the system is subject to breakdowns, say at an exponential rate β with an exponential repair time at rate δ. When the system is down, a job attempting to leave the system has to remain at the station from which it came.

The present system has *no closed product-form* expression for its steady-state queue length distribution, because when the system channel is down:

the system outflow $= 0$ but the system inflow is still > 0.

Product-form modification To obtain a computationally simple performance bound for the described system the following modification could therefore be suggested:

when the departure channel is down also reject arrivals.

Under this modification the following product-form expression is readily verified for any state $(\mathbf{n}, \theta)$ where $\mathbf{n} = (n_1, ..., n_N)$ represents the population vector of jobs present, and where $\theta = 1$ if the channel is up and $\theta = 0$ if the channel is down:

$$\bar{\pi}(\mathbf{n}, \theta) = c \prod_i \prod_{k=1}^{n_i} \left[\frac{\lambda_i}{\mu_i(k)} \right] \left\{ 1_{\{\theta=1\}} + 1_{\{\theta=0\}} \left[\frac{\beta}{\delta} \right] \right\}. \tag{6.44}$$

Here c is a normalizing constant, $1_{\{A\}} = 1$ if condition A is satisfied and $1_{\{A\}} = 0$ if not, and the values λ_i are the throughputs as determined by the traffic equations $\lambda_i = \lambda p_{0i} + \sum_j \lambda_j p_{ji}$. With $n = n_1 + \cdots + n_N$ denoting the total number of jobs present, the expected loss fraction is now readily evaluated as:

$$
\begin{aligned}
\bar{\mathbf{B}} \;=\; & \text{fraction of jobs lost due to a capacity limit} \\
& +\text{fraction of jobs lost when the system is down} \\
=\; & \sum_{\{\mathbf{n}|n=F\}} \pi(\mathbf{n},1) + \sum_{\{(\mathbf{n},\theta)|\theta=0,n<F\}} \pi(\mathbf{n},0).
\end{aligned}
\tag{6.45}
$$

By proving that $\bar{\mathbf{B}}$ is an upper bound (denoted by $\mathbf{B}_U$) for the original loss probability and similarly by claiming that $\bar{\mathbf{B}}$ would be a lower bound (denoted by $\mathbf{B}_L$) if we would assume that breakdowns do not take place, i.e., by substituting $\beta = 0$, we would thus have obtained relatively simple expressions to capture the real loss probability $\mathbf{B}$ (or throughput $\mathbf{T} = \lambda(1 - \mathbf{B})$) between a lower and upper bound value (or conversely an upper and lower bound). Also assuming that the down time ratio β/δ is rather small, as is usual, these bounds can be quite accurate (as shown below):

$$
\mathbf{B}_L \le \mathbf{B} \le \mathbf{B}_U
\tag{6.46}
$$

Upper bound $\mathbf{B}_U$ To prove the inequality $\mathbf{B} \le \mathbf{B}_U$ in the setting of Section 6.4.2, let the modified product-form system be denoted with an overbar symbol, and identify a state i as $(\mathbf{n}, \theta)$, and a state j as $(\mathbf{n}, \theta)'$. In order to count the expected number of losses in the original and in the modified model, we choose the reward rates:

$$
\begin{cases}
r(\mathbf{n},\theta) &= \lambda 1_{\{n=F\}}, \\
\bar{r}(\mathbf{n},\theta) &= \lambda 1_{\{n=F\}} = \lambda 1_{\{\theta=0,n<F\}}.
\end{cases}
\tag{6.47}
$$

By comparing the transition rates $\bar{q}(i,j)$ and $q(i,j)$, and with $\mathbf{n} + e_i$ the vector equal to $\mathbf{n}$ with one job more at station i, we obtain:

$$
\sum_{(\mathbf{n},\theta)} [\bar{q}((\mathbf{n},\theta),(\mathbf{n},\theta)') - q((\mathbf{n},\theta),(\mathbf{n},\theta)')] \, [\mathbf{V}^k(\mathbf{n},\theta)' - \mathbf{V}^k((\mathbf{n},\theta))]
$$
$$
= \lambda 1_{\{\theta=0,n<F\}} \sum_i p_{0,i} \, [\mathbf{V}^k(\mathbf{n},0) - \mathbf{V}^k((\mathbf{n}+e_i,0)].
\tag{6.48}
$$

By (6.47), (6.48) and the essential Lemma 6.2 below, we can obtain:

$$[\bar{r}(\mathbf{n},\theta) - r(\mathbf{n},\theta)] +$$

$$\sum_{(\mathbf{n},\theta)'} [\bar{q}((\mathbf{n},\theta),(\mathbf{n},\theta)') - \mathbf{q}((\mathbf{n},\theta),(\mathbf{n},\theta)')][\mathbf{V}^k(\mathbf{n},\theta)' - \mathbf{V}^k((\mathbf{n},\theta))] \geq 0.$$

(6.49)

With $\bar{\mathbf{G}} = \lambda\bar{\mathbf{B}}_U$ and $\bar{\mathbf{G}} = \lambda\bar{\mathbf{B}}$, Theorem 6.2 thus implies that $\mathbf{B}_U \geq \mathbf{B}$. Similarly, we prove $\mathbf{B}_L \leq \mathbf{B}$, from which (6.46) follows.

Remark 3: Non-monotonicity Note here that the reward rate term $[\bar{r}(\mathbf{n},\theta) - r(\mathbf{n} - \theta)]$ is required in (6.49) as the second term does lead to a negative contribution due to (6.48). This in fact implies that the present application is *not monotone* and *did require* the Markov reward approach.

Discussion counter-example and formal proof completion Essentially, the Markov reward approach overcomes this non-monotonicity feature by using *expectations* instead of sample path relations, as stochastic comparison methods are directly related to. To shed some more light on this non-monotonicity feature, below we give a *counterintuitive* example. This example shows that on a *realization basis*, the rejection of jobs when the system is down may actually increase the throughput. Essential for the negative terms appearing in (6.49), which can be regarded as *non-monotonicity*, to be sufficiently compensated by reward terms, is the following technical lemma, where not only a lower estimate but also an upper estimate is obtained for the difference terms as required in (6.47)–(6.49).

Lemma 6.2 *For all* $(\mathbf{n},\theta)$, i *and* $k \geq 0$:

$$0 \leq \mathbf{V}^k(\mathbf{n} + e_i, \theta) - \mathbf{V}^k(\mathbf{n},\theta) \leq 1. \tag{6.50}$$

Proof. For presentational convenience and instructive purposes we restrict the proof to the case of a single queue with service rate $\mu(n)$ when n jobs are present. The essence of blocking and breakdowns is hereby still covered while the extension to a queueing network is purely notational, as for example in [443] without breakdowns and blocking.

The proof follows by induction to k. For $k = 0$, (6.50) applies, as $V^0(.) = 0$. Assume that (6.50) holds for $k = m$. Then, in a similar fashion as in the proof of Lemma 6.1, after some substitutions and rewriting we can derive:

$$\mathbf{V}^{m+1}(n+1,\theta) - \mathbf{V}^{m+1}(n,\theta)$$

$$= \lambda 1_{\{n+1=F\}}\Lambda^{-1} + \lambda\Lambda^{-1}1_{\{n+1=F\}}[\mathbf{V}^m(n+1,\theta) - \mathbf{V}^m(n+1,\theta)]$$

$$+ \lambda\Lambda^{-1}1_{\{n+1=F\}}[\mathbf{V}^m(n+2,\theta) - \mathbf{V}^m(n+1,\theta)]$$

$$+ \mu(n)\Lambda^{-1}1_{\{\theta=1\}}[\mathbf{V}^m(n,\theta) - \mathbf{V}^m(n-1,\theta)]$$

$$+ \mu(n)\Lambda^{-1}1_{\{\theta=0\}}[\mathbf{V}^m(n+1,\theta) - \mathbf{V}^m(n,\theta)]$$

$$+ [\mu(n+1) - \mu(n)]\Lambda^{-1}1_{\{\theta=1\}}[\mathbf{V}^m(n,\theta) - \mathbf{V}^m(n,\theta)]$$

$$+ [1 - \lambda\Lambda^{-1} - \mu(n+1)\Lambda^{-1}1_{\{\theta=1\}}][\mathbf{V}^m(n+1,\theta) - \mathbf{V}^m(n,\theta)].(6.51)$$

The second to last and preultimate terms in the right-hand side are indeed equal to 0 but kept in for clarity of presentation. Further, note that we must have assumed, as per (6.1), that $\Lambda \geq \lambda + \mu(n+1)$, for all $n+1 \leq \mathbf{F}$. As a consequence, by substituting the lower estimate (i.e., 0) from (6.50) with $k = m$, one directly concludes that $\mathbf{V}^{m+1}(n+1,\theta) - \mathbf{V}^{m+1}(n,\theta) \geq 0$. By substituting the upper estimate (i.e., 1) from (6.50) with $k = m$ and observing that the second term compensates for the first additional term, we also conclude that the right-hand side of (6.51) is estimated from above by 1. The induction now completes the proof. □

A numerical illustration In Table 6.1 we give some numerical values for the simple case of a single queue subject to breakdowns, as adopted from [118]. Also here no simple closed-form expression is available for **B**. Read $\rho = \lambda/\mu$ and $\tau = \beta/\delta$. Here τ can be seen as approximately the fraction of time for which the system is down, which should realistically be thought of as being rather small, say in the order of 2% ($\tau = 0.02$). In such cases the results show that the lower and upper bounds $\mathbf{B}_L$ and $\mathbf{B}_U$

Table 6.1: Single lower and upper bounds for the loss fraction **B**

F	ρ	τ	$\mathbf{B}_L$	$\mathbf{B}_U$
20	20	0.1	0.16	0.24
		0.05	0.16	0.20
		0.02	0.16	0.18
30	25	0.05	0.052	0.098
		0.01	0.052	0.062
20	15	0.05	0.045	0.091
		0.01	0.045	0.065
		0.005	0.045	0.055
10	5	0.01	0.018	0.028
		0.005	0.018	0.024
		0.001	0.018	0.020

even provide quite reasonable and guaranteed estimates of the real value **B** as per (6.46).

6.5   Error bound results

Next to the advantage of the Markov reward approach for comparison purposes in specific situations, the major advantage of this approach is its potential to also provide *error bounds* when approximating a CTMC with a slightly modified version. As this has already been discussed and illustrated in depth in various earlier papers, most notably [119, 121, 122, 131], the presentation below will be restricted to a compact form with an extension to steady-state weighting and will be tailored to two performability applications. First, Section 6.5.1 provides the general theorem and a brief discussion of its twofold nature for usage. Next, Section 6.5.2 gives two performability applications.

6.5.1   General error bound theorem

Reconsider the setting of Section 6.4.2 with:

$$\begin{cases} \text{an original Markov reward chain } (S, \mathbf{q}, r), \\ \text{an approximate Markov reward chain } (\bar{S}, \bar{\mathbf{q}}, \bar{r}), \end{cases}$$

where both are assumed to be uniformizable with the same constant Λ and where $\bar{S} \subset S$. Let π and $\bar{\pi}$ denote their steady-state distributions. The following theorem can be given in various versions as presented earlier (notably [121]). The present form, however, is slightly more convenient if the steady-state distribution of one of the two chains, say the second, is known as easily computable. For convenience write $\bar{\pi} f = \sum_i \bar{\pi}(i) f(i)$.

Theorem 6.3 (Error bound) *Suppose that for some function $\gamma(.)$ at $\bar{S}$ all $i \in \bar{S}$ and $k \geq 0$:*

$$[\bar{r} - r](i) + \sum_j [\bar{\mathbf{q}}(i,j) - \mathbf{q}(i,j)][\mathbf{V}^k(j) - \mathbf{V}^k(i)] \leq \gamma(i), \qquad (6.52)$$

then

$$|\bar{G} - G| \leq \sum_i \bar{\pi}(i)\gamma(i) = \bar{\pi}\bar{\gamma}. \qquad (6.53)$$

Proof. Recall the derivation (6.40) for fixed $\ell \in \bar{S}$. Then multiplication by $\bar{\pi}(\ell)$ and summing over all ℓ we obtain:

$$\bar{\pi}\bar{\mathbf{V}}^k - \bar{\pi}\mathbf{V}^k = \bar{\pi}\sum_{s=0}^{k-1} \bar{\mathbf{T}}^s \left\{ [\bar{r} - r]\Lambda^{-1} + [(\bar{\mathbf{T}} - \mathbf{T})\mathbf{V}^{k-s-1}] \right\}$$

$$= \sum_{s=0}^{k-1} (\bar{\pi} \left\{ [\bar{r} - r]\Lambda^{-1} + [(\bar{\mathbf{T}} - \mathbf{T})\mathbf{V}^{k-s-1}] \right\}, \qquad (6.54)$$

where we used the fact that

$$\bar{\pi}\bar{T}^s f = \bar{\pi}\bar{T}(\bar{T}^{s-1} f) = \bar{\pi}(\bar{T}^{s-1} f) = \cdots = \bar{\pi} f \qquad (6.55)$$

since $\bar{\pi}$ is invariable under $\bar{T}$ (steady-state measure). As a consequence,

$$b|\bar{\pi}\bar{\mathbf{V}}^k - \bar{\pi}\mathbf{V}^k| \leq k \sum_i \bar{\pi}(i) \frac{[\bar{r} - r](i)}{\Lambda} = (\bar{T} - T)\mathbf{V}^{k-s-1}(i)|. \qquad (6.56)$$

Substitution of (6.41) in (6.56) and using condition (6.52) thus gives

$$|\bar{\pi}\bar{\mathbf{V}}^k - \bar{\pi}\mathbf{V}^k| \le k\Lambda^{-1}\sum_i \bar{\pi}(i)\gamma(i) = k\Lambda^{-1}[\bar{\pi}\gamma]. \qquad (6.57)$$

By recalling the steady state convergence (6.8) to be independent of the initial state, the proof is thus completed by:

$$
\begin{cases}
\frac{\Lambda}{k}\sum_i \bar{\pi}(i)\bar{\mathbf{V}}^k(i) & \to \quad \bar{\mathbf{G}}, \qquad k \to \infty, \\[2mm]
\frac{\Lambda}{k}\sum_i \bar{\pi}(i)\mathbf{V}^k(i) & \to \quad \mathbf{G}, \qquad k \to \infty.
\end{cases} \qquad (6.58)
$$

$\square$

Discussion of Theorem 6.3 First of all, in analogy with Theorem 6.2, Theorem 6.3 also relies essentially upon being able to estimate (find bounds for) the bias difference terms $\mathbf{V}^k(i) - \mathbf{V}^k(j)$, where i and j only need to be considered as one-step neighbours. In specific applications, a technical lemma, such as Lemma 6.2, will thus also be required.

Next, it is worthwhile to note that Theorem 6.3 may lead to small error bounds in either of two ways:

1. when either the difference between the transition rates $\mathbf{q}$ and $\bar{\mathbf{q}}$ is small, uniformly in all states, where one may typically think of small *perturbations or inaccuracies* in system parameters such as an arrival rate λ, or

2. when the transition rates $\mathbf{q}$ and $\bar{\mathbf{q}}$ may differ quite strongly in specific states i, but where the likelihood $\bar{\pi}(i)$ og being in such states is rather small. Here one could typically think of a system *truncation or modification*.

The applications in Section 6.5.2 belong to these cases, respectively.

6.5.2 Two special performability applications

Availability model Reconsider the performability model from Section 6.3.1 under conditions (6.19) and (6.20). Now define a truncated version in which at most one component can be down. In other words, we may substitute

$$\bar{\beta}(h|H) = \begin{cases} \beta_h > 0, & \text{if } H = \emptyset, \\ 0, & \text{otherwise,} \end{cases} \qquad (6.59)$$

$$\bar{\delta}(h|h) = \delta_n. \tag{6.60}$$

Then

$$\bar{\pi}(h) = c \left[\frac{\beta_h}{\delta_h}\right], \qquad h \in \{1, ..., N\}, \tag{6.61}$$

presents the steady-state distribution of the truncated version, while for the original version a product form is available only under special conditions (recall cases 1 and 2 in Section 6.2.3).

Let $\bar{r}(H) = r(H) = [n|N]$ to measure the down fraction of components. Hence, $[\bar{r} - r](.) = 0$. Further, with H identified with a state i and H' as a state j, we have for $H = \{h\} \in \bar{S}$:

$$\sum_{H'}[\bar{q}(H, H') - q(H, H')][\mathbf{V}^k(H') - \mathbf{V}^k(H)] = \\ \sum_{s \neq h} \beta(s|h)[\mathbf{V}^k(h+s) - \mathbf{V}^k(h)]. \tag{6.62}$$

Furthermore, in a similar fashion as in Lemmas 6.1 and 6.2, it can be proven that for all $H, H + h$:

$$0 \leq \mathbf{V}^k(H + h) - \mathbf{V}(H) \leq 1 \tag{6.63}$$

Consequently, by Theorem 6.3, relations (6.62) and (6.63), we have:

$$|\bar{G} - G| \leq c \sum_h \frac{\beta_h}{\delta_h} \left[\sum_{s \neq h} \beta(s|h)\right], \tag{6.64}$$

$$\bar{G} = c \sum_h \frac{\beta_h}{\delta_h}, \qquad c = \left[1 + \sum_h \frac{\beta_h}{\delta_h}\right]^{-1},$$

where we have assumed the scaling:

$$\sum_{s \notin H} \beta(s|H) + \sum_{h \in H} \delta(h|H) \leq 1, \tag{6.65}$$

for all H. In other words, the error in the expected number (fraction) of down components using the simple expression (6.61) is of the order of having two or more components down. As this is usually relatively small, the proposed simplification can be useful for quick approximative purposes. Other error bound results for performability applications can be derived along similar lines but are left to the reader.

A telecommunication example To illustrate how the results of this chapter apply to servicing or communication networks, we briefly present an application to telecommunications.

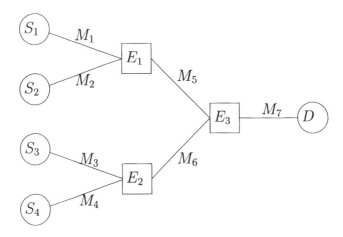

Figure 6.5: A telecommunication example

Consider a circuit-switched communication network as shown in Figure 6.5, with type i call request at a Poisson arrival rate λ_i and exponential call (or holding) times with parameter μ_i. A call requires a free link (trunk or circuit) from each of the linkgroups along its route. For example, a type 1 call requires one link from linkgroup 1, linkgroup 5 and linkgroup 7 to be available. When no available path can be found a call request is lost. Let $\mathbf{M}_i$ be the number of links of linkgroup i, $i = 1, \cdots, 7$. Then, by verifying the balance equations per type, one easily proves that the steady-state distribution of the state $\mathbf{n} = (n_1, n_2, n_3, n_4)$, with n_i the number of ongoing type i calls, exhibits the product form:

$$\pi(\mathbf{n}) = c \prod_i \frac{1}{n_i} \left[\frac{\lambda_i}{\mu_i}\right]^{n_i}, \quad \begin{cases} n_i \le M_i & i = 1, \cdots, 2, \\ n_1 + n_2 \le M_5, \\ n_3 + n_4 \le M_6, \\ n_1 + n_2 + n_3 + n_4 \le M_7. \end{cases} \quad (6.66)$$

Now assume that one of the linkgroups is subject to breakdowns, say linkgroup 5, at an exponential rate δ. When a linkgroup goes down, all calls that use a link from that group are lost. A product-form expression then no longer applies, nor can a product-form modification as in Section 6.4 be suggested , since interrupted calls are lost and not resumed. However, by assuming that the downtime ratio $\gamma = \beta/\delta$ is small, say less than 1%, it is appealing to use the above product form as an approximation as if breakdowns are ignored. We would like to know the (in)accuracy of this approximation for specific measures, say the throughput of successfully completed calls per unit time of a specific type of call, e.g., type 2. To this end, we use state (n, θ), where we also denote by θ the status of this special linkgroup 5 to be up $(\theta = 1)$ or down $(\theta = 0)$. Then with $(\mathbf{n}, \theta)$ identified with a state i and $(\bar{\mathbf{n}}, \theta)$ as a state j we obtain:

$$
\sum_{(\mathbf{n},\theta)'} [\bar{\mathbf{q}}((\mathbf{n}, \theta), (\mathbf{n}, \theta)') - \mathbf{q}((\mathbf{n}, \theta), (\mathbf{n}, \theta)')][\mathbf{V}^k((\mathbf{n}, \theta)') - \mathbf{V}^k((\mathbf{n}, \theta))] =
$$
$$
1_{\{\theta=1\}}\beta[\mathbf{V}^k((\mathbf{n}, 0)) - \mathbf{V}^k((\mathbf{n} - L_5, 0))]+
$$
$$
1_{\{\theta=1\}}\lambda_1 1_{\{n+e_1+\in C\}}[\mathbf{V}^k((\mathbf{n} + e_1, 0)) - \mathbf{V}^k((\mathbf{n}, 0))]+
$$
$$
1_{\{\theta=1\}}\lambda_2 1_{\{n+e_2+\in C\}}[\mathbf{V}^k((\mathbf{n} + e_2, 0)) - \mathbf{V}^k((\mathbf{n}, 0))],
$$

$$(6.67)$$

where $\mathbf{n} - L_5$ represents the state equal to $\bar{n}$ with all calls using linkgroup 5 deleted and where C is the set of admissible states as indicated above. Furthermore by choosing

$$
\bar{r}(\mathbf{n}, \theta) = r(\mathbf{n}, \theta) = n_2\mu_2 \tag{6.68}
$$

to measure the throughput of type 2 calls, we have $\bar{r} = r$ and we can prove similarly to Lemma 6.1, that for $\theta = 0, 1$ and $i = 1, \cdots, 4$:

$$
0 \le \mathbf{V}^k((\mathbf{n}, \theta)) - \mathbf{V}^k((\mathbf{n} - L_5, \theta)) \le [n_1 + n_2]. \tag{6.69}
$$

By Theorem 6.3 and relations (6.67)–(6.69), we can thus conclude the following error bound for the throughput of type 2 calls calculated by (6.66) and (6.68) ($\bar{\mathbf{G}}$), and of the original model with breakdowns ($\mathbf{G}$):

$$
|\bar{\mathbf{G}} - \mathbf{G}| \le \{\sum_n \bar{\pi}(n)[n_1 + n_2] + [\lambda_1 + \lambda_2]\}\bar{\tau}, \tag{6.70}
$$

where $\bar{\tau} = \beta/\delta$, with δ scaled to 1, can be seen as the fraction of time for which linkgroup 5 is down.

6.6   Summary

In this chapter, it was shown how Markov reward structures can be used for performability analysis to establish:

1. *monotonicity results* of expected performance measures, such as availability, over *time*,

2. *comparison results* of steady-state performability measures when comparing two systems, where one system is typically a modification of the other,

3. *error bound results* for situations as under 2, either where the modification is small or where it will only take place with small probability.

The verification of the necessary conditions appears feasible in an analytic manner in applications which themselves are not analytically solvable. As such, the approach outlined can be most helpful to *support* practical engineering in performability.

Chapter 7

The Task Completion Time in Degradable Systems

Andrea Bobbio
Miklós Telek

This chapter assumes a user-oriented point of view in examining the performability of a dependable computing system. The investigated performability measure is the effective time that a task, with an assigned work requirement, takes to be executed by the system. Assuming that the system changes its performance characteristics randomly in time, the stochastic model representing the task completion time is formulated and analyzed. Applications and extensions of the basic model are discussed. Finally, the completion time model is reformulated in the language of stochastic Petri nets, and possible computational approaches are illustrated.

7.1 Introduction

THE completion time of a task measures the time that a task takes to be executed by a computing system. If the system changes its computational power randomly in time during the execution, the task completion time is a random variable. The analytical and numerical computation of the cumulative distribution function (CDF) of the task completion time is the subject of this chapter.

The adopted modelling framework consists of describing the behaviour of the system configuration in time by means of a stochastic process, called the *structure-state process*, and by associating to each state of the structure-state process a non-negative real constant representing the effective working capacity or performance level of the system in that state. The real variable associated to each structure-state is called the *reward rate* [236]. The structure-state process together with the reward rates forms the *stochastic reward model* (SRM) [384].

The properties of stochastic reward processes have been studied a long time [303, 88, 259, 260, 236]; however, only recently have SRMs received attention as a modelling tool in performance/reliability evaluation. Indeed, the possibility of associating a reward variable to each structure state increases the descriptive power and the flexibility of the model.

Different interpretations of the structure-state process and of the associated reward structure give rise to different applications [319]. Common assignments of the reward rates are execution rates of tasks in computing systems (the computational capacity) [27, 450], number of active processors (or processing power) [39, 188], throughput [316, 154, 201], average response time [237, 264, 283], and response time distribution [464, 383, 463]. The classical reliability theory [22] can be viewed as a particular case of SRM obtained by constraining the reward rates to be binary variables.

Two main different points of view have been assumed in the literature when dealing with SRM for degradable systems [271]. In the *system-oriented* point of view the most significant measure is the total amount of work done by the system in a finite interval. The accumulated reward is a random variable whose CDF is called the *performability* [315]. Various numerical techniques for the evaluation of the performability have appeared in the literature [240, 136, 186, 440, 104, 370, 371, 108]. In the *user-oriented* (or *task-oriented*) point of view the system is regarded as a server, and the emphasis of the analysis is on the ability of the system to provide a prescribed service in due time. Consequently, the most characterizing measure becomes the probability of accomplishing an assigned task in a given time. The task-oriented point of view is a more direct representation of the quality of service, which, in turn, is the main target of a dependable computation.

Gaver [153] analyzed the distribution of the completion time for a two-state server with different mechanisms of interruption and recovery policies. Extensions to the above model were provided in [354], whereas the completion time problem for fault tolerant computing systems was addressed in [65]. A unified formulation to the system-oriented and the user-oriented point of view was provided by Kulkarni, Nicola and Trivedi in [271, 272, 356]. An alternative interpretation of the completion time problem can be given in terms of the hitting time of an appropriate cumulative functional [88] against an absorbing barrier equal to the work requirement. The definition of a cumulative functional was first suggested in [271] and then explicitly exploited in [44], where the completion time was modelled as a first hitting time against an absorbing barrier. This interpretation leads the above problem into the mainstream of absorption problems in stochastic models and has proved to be useful in association with stochastic Petri nets [40] and with the extension to multi-reward models [41, 44].

In Section 7.2, the completion time problem is formulated as a first passage time across an absorbing barrier. The distribution of the completion time is derived in Section 7.3, in the Laplace transform domain and under the hypothesis that all the states pertain to the same preemption class. Section 7.4 illustrates some applications and extensions. Section 7.5 shows how to represent the formulated non-Markovian stochastic model by means of Petri nets, and compares the results obtained from two non-Markovian Petri net-based models on a simple example. Section 7.6 summarizes the chapter.

7.2   The barrier hitting problem

Given that $F(t)$ is a CDF, the Laplace transform (LT) $F^*(s)$ and the Laplace–Stieltjes transform (LST) $F^\sim(s)$ are given by, respectively:

$$F^*(s) = \int_0^\infty e^{-st} F(t)\, dt, \qquad F^\sim(s) = \int_0^\infty e^{-st}\, d\, F(t).$$

Let the structure-state process $(X(t), \; t \geq 0)$ be a right-continuous semi-Markov process [88, 269] defined over a discrete and finite state space $\mathcal{S}$ of cardinality n. We denote by H the time duration until the first embedded time point of the semi-Markov process starting from state i at time 0 $(X(0) = i$), and by $\mathbf{p}(0)$ the row vector of the initial probabilities. Let $\mathbf{K}(t) = [K_{ij}(t)]$ be the kernel of the semi-Markov process. The generic element

$$K_{ij}(t) = \Pr\{H \leq t, X(H) = j | X(0) = i\},$$

with $i, j = 1, ..., n$, is the distribution of H starting in state i at time 0 supposing that a transition to state j took place. Moreover,

$$K_i(t) = \Pr\{H \leq t | X(0) = i\} = \sum_{j=1}^{n} K_{ij}(t), \qquad i = 1, ..., n,$$

is the distribution of H starting in state i at time 0 independent of the state reached after the first embedded time point. The probability of jumping from state i to state j at time $H = t$ can be defined in terms of the kernel elements:

$$\Pr\{X(H) = j | H = t, X(0) = i\} = \frac{dK_{ij}(t)}{dK_i(t)}.$$

Let $r_{X(t)}$ be a non-negative real-valued function defined as:

$$r_{X(t)} = r_i \quad \text{if} \quad X(t) = i \quad \text{with} \quad r_i \geq 0 \quad \text{and} \quad i = 1, 2, \ldots, n. \quad (7.1)$$

$r_{X(t)}$ represents the instantaneous reward associated to state i. We now define a functional $Y(t)$ that represents the accumulation of reward in time. $Y(t)$ is a stochastic process that depends on $X(\tau)$ for $\tau \leq t$ [88]. During the sojourn of $X(t)$ in state i between t and $t + dt$, $Y(t)$ increases at the rate $r_i \, dt$. However, a transition in $X(t)$ may induce a modification in the accumulation process depending on whether the transition entails a *loss of work*, or *no loss of work*. A transition which does not entail a loss of the work already accumulated by the system on the task in execution is called *preemptive resume (prs)*, and its effect on the model is that the functional $Y(t)$ resumes the previous value in the new state. A transition which entails a loss of the work done by the system on the task in execution is called *preemptive repeat (prt)*, and its effect on the model is that the functional $Y(t)$ is reset to 0 in the new state.

A possible realization of the accumulation process $Y(t)$ is shown in Figure 7.1. The transition from state j to state k is of *prs* type, whereas the transitions from state k to i and from i to j are *prt*. In order to model the completion time problem, let W be the actual work requirement of a task. W represents the time required to execute a task in isolation on a perfect system. In a degradable environment, the task completes as soon as the work accumulated by the system reaches the actual work requirement for the first time. Hence, W acts as an absorbing barrier for the functional $Y(t)$. With reference to Figure 7.1, the task completion time is the time at which $Y(t)$ hits the barrier W for the first time.

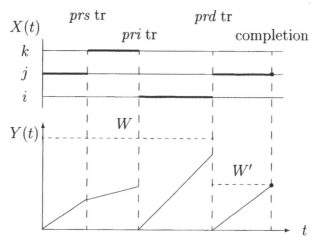

Figure 7.1: The behaviour of the stochastic process $X(t)$ and the functional $Y(t)$ versus time t

We assume, in general, that W is a random variable with distribution $G(x)$ with support on $[0, \infty)$. The degenerate case in which W is deterministic and the distribution $G(x)$ becomes the unit step function $U(x)$ located at $W = x$, can be considered as well. When W is not deterministic and the preemption policy is *prt*, two cases arise depending on whether the repeated task has an identical work requirement to the original preempted task: *preemptive repeat identical (pri)*, or a different work requirement sampled from the same distribution: *preemptive repeat different (prd)*. With reference to Figure 7.1, it is assumed that the transition from state k to state i is *pri* and the transition from state i to state j is *prd*. According to the previous assumptions, the accumulated reward $Y(t)$ is reset and the same value W of the barrier is retained when jumping from state k to state i since the corresponding transition is *pri*. On the other hand, since transition from i to j is *prd*, the work requirement W is resampled in state j assuming a new value W' sampled from the same distribution, and $Y(t)$ is reset.

For a structure-state process with only *prs* and *pri* transitions the barrier height W is constant up to the completion. In these cases, conditioned to a fixed value of the barrier height $W = x$, the completion time $T(x)$ is defined as:

$$T(x) = \min [t \geq 0 : Y(t) = x]. \tag{7.2}$$

Let $F_T(t, x)$ be the conditional CDF of the task completion time $T(x)$:

$$F_T(t, x) = \Pr\{T(x) \leq t\}. \tag{7.3}$$

The unconditional completion time T is characterized by the following distribution:

$$F_T(t) = \Pr\{T \leq t\} = \int_0^\infty F_T(t, x) \, dG(x), \tag{7.4}$$

and is the measure that can be evaluated if all the transitions are *prd*.

The distribution of the completion time, $F_T(t)$, incorporates the effect of a random variation of the execution speed consequent to a degradation and reconfiguration process, combined with the effect of the preemption and recovery policy on the execution of the task.

The following relationships between the different preemption policies can be easily established. If the work requirement W is an exponential random variable, the two policies *prs* and *prd* give rise to the same completion time (due to the memoryless property of the exponential distribution, the residual task requirement under the *prs* policy coincides with the resampled requirement under the *prd* policy). On the other hand, if W is deterministic, the two policies *pri* and *prd* are coincident (resampling a step function provides always the same constant value).

Moreover, assuming that the structure-states are all of *prs* type, so that no loss of reward occurs,

$$Y(t) = \int_0^t r_{X(\tau)} \, d\tau$$

and the distribution of the completion time is closely related to the distribution of the accumulated reward (*performability*) by means of the following relation:

$$\Pr\{Y(t) \leq x\} = \Pr\{T(x) \geq t\}. \tag{7.5}$$

Closed-form Laplace transform equations for $F_T(x, t)$ when $X(t)$ is a CTMC and all the states belong to the same preemption class were derived in [271]. The extension to a semi-Markov process $X(t)$ whose state space is partitioned in the three preemption classes has been considered in [272]. Bobbio and Trivedi [48] studied the case where $X(t)$ is a CTMC, the work requirement W is a phase-type (PH) random variable [350] and the task execution policy is a probabilistic mixture of *prs* and *prd* policies. The combination of *prs* and *pri* policies has been investigated in [70] for the evaluation of the completion time of a program on a gracefully-degradable computing system.

7.3 The distribution of the completion time

A state whose outgoing transitions are all of *prs* type is called a *prs* state; similarly, a state whose outgoing transitions are all of *prd* (*pri*) type is called a *prd* (*pri*) state. The following closed-form expressions for the CDF of the completion time are derived under the hypothesis that the structure-state process is semi-Markov and all the states are of the same preemption class. A more general derivation in which the states are allowed to belong to the three different defined preemption classes can be found in [272].

In order to evaluate (7.3), let us introduce the following vector valued functions $\mathbf{F}(t)$ and $\mathbf{F}(t, x)$ whose entries $F_i(t)$ and $F_i(t, x)$, $i = 1, 2, \ldots, n$, are defined by:

$$F_i(t, x) = \Pr\{T(x) \leq t \mid X(0) = i\}, \quad x \geq 0 \tag{7.6}$$

$$F_i(t) = \Pr\{T \leq t \mid X(0) = i\}. \tag{7.7}$$

Notice that, when all the states are *prs* or *pri* the quantity to be evaluated is $F_i(t, x)$ while in the *prd* case only the function $F_i(t)$ can be derived. From the above definitions, it follows that:

$$F_T^\sim(s) = \mathbf{p}(0) \, \mathbf{F}^\sim(s) = \int_0^\infty \mathbf{p}(0) \, \mathbf{F}^\sim(s, x) \, dG(x). \tag{7.8}$$

Theorem 7.1 *Given that $X(t)$ is a semi-Markov process and all the states are of prs type, the double transform $\mathbf{F}^{\sim*}(s, w)$ satisfies the following equation:*

$$F_i^{\sim*}(s, w) = \frac{r_i}{s + w\, r_i}\, [1 - K_i^{\sim}(s + w\, r_i)] + \sum_{j=1}^{n} K_{ij}^{\sim}(s + w\, r_i)\, F_j^{\sim*}(s, w).$$

(7.9)

Proof. Conditioning on the time until the first embedded time point in the initial state $H = h$, let us define:

$$F_i^{\sim}(s, x \mid H = h) = E[exp(-s\, T)|W = x, X(0) = i, H = h]$$

$$= \begin{cases} exp(-sx/r_i), & \text{if } h\, r_i \geq x, \\ exp(-sh) \sum_{j=1}^{n} \dfrac{dK_{ij}(h)}{dK_i(h)} F_j^{\sim}(s, x - h r_i), & \text{if } x > h r_i. \end{cases}$$

(7.10)

In (7.10), two mutually exclusive events are identified: if $h\, r_i \geq x$, then $T = x/r_i$ or if $h\, r_i < x$ then a transition occurs to state j. Taking the LT with respect to x, we obtain:

$$F_i^{\sim*}(s, w|H = h) = \int_{x=0}^{hr_i} exp(-wx)exp(-sx/r_i)dx$$

$$+ \; exp[-(s + wr_i)h] \sum_{j=1}^{n} \frac{dK_{ij}(h)}{dK_i(h)} F_j^{\sim*}(s, w). \, (7.11)$$

Unconditioning with respect to H, (7.11) becomes:

$$F_i^{\sim*}(s, w) = \int_{h=0}^{\infty} \int_{x=0}^{hr_i} exp[-(s + w\, r_i)x/r_i]dx\, dK_i(h)$$

$$+ \int_{h=0}^{\infty} exp[-(s + w\, r_i)h] \sum_{j=1}^{n} F_j^{\sim*}(s, w)\, dK_{ij}(h). \, (7.12)$$

Finally, (7.9) is obtained from (7.12) by evaluating the integrals.   □

Corollary 7.1 *Under the assumptions of Theorem 7.1, given that $X(t)$ is a CTMC with infinitesimal generator $\mathbf{Q}$, the double transform $\mathbf{F}^{\sim*}(s, w)$ satisfies the following matrix equation (see [271]):*

$$\mathbf{F}^{\sim*}(s, w) = [s\mathbf{I} + w\mathbf{R} - \mathbf{Q}]^{-1}\, \mathbf{r},$$

(7.13)

where

$$\mathbf{r} = [r_1, r_2, \ldots, r_n]^T; \qquad \mathbf{R} = diag\,[r_1, r_2, \ldots, r_n] \qquad (7.14)$$

are a vector and a matrix of reward rates, respectively.

Proof. Equation (7.13) is obtained by substituting the following Markovian kernel in (7.9):

$$K_{ij}^{\sim}(s) = \begin{cases} \dfrac{q_{ij}}{s + q_i}, & \text{if } i \neq j, \\ 0, & \text{if } i = j, \end{cases}$$

where $q_i = \sum_{j=1; j \neq i}^{n} q_{ij}$. □

In general, the kernel $\mathbf{K}(t)$ of a semi-Markov process can have non-zero positive entries on the main diagonal. Therefore, from a given state i either a "virtual" transition into state i itself or a real transition to a different state $j \neq i$ can take place. In the previously considered *prs* case, the accumulation process resumes the value reached by the total reward in the previous state and there is no need to distinguish between a jump into the same state or into a different one.

The situation is different in the *prt* case (either *prd* or *pri*). Indeed, a jump into a new state resets the accumulated reward, whereas a jump into the same state should retain the same reward level. However, it has been shown in [458] that a semi-Markov kernel can be transformed into a canonical form in which the entries on the main diagonal are zero while preserving the same transition probabilities for all the transitions from i to j with $i \neq j$. The canonical representation of the semi-Markov kernel $\mathbf{K}^u(t)$ is given by [458]:

$$K_{ij}^{u\sim}(s) = \begin{cases} \dfrac{K_{ij}^{\sim}(s)}{1 - K_{ii}^{\sim}(s)}, & \text{if } i \neq j, \\ 0, & \text{if } i = j. \end{cases} \qquad (7.15)$$

With a kernel in canonical form, the problem of distinguishing between transitions into the same state or into a different state is avoided. Therefore, in the following we implicitly assume that the kernel is, or has been, transformed in canonical form with zero entries on the main diagonal.

Theorem 7.2 *Given that $X(t)$ is a semi-Markov process and all the states are of prd type, the LST $F_i^\sim(s)$ satisfies the following equation:*

$$F_i^\sim(s) = \int_0^\infty exp(-s\,x/r_i)\,[1 - K_i(x/r_i)]\,dG(x)$$

$$+ \sum_{j=1}^n F_j^\sim(s) \int_0^\infty exp(-sh)\,[1 - G(hr_i)]\,dK_{ij}(h). \quad (7.16)$$

Proof. Conditioning on the time until the first embedded time point in the initial state $H = h$, let us define:

$$F_i^\sim(s\,|\,W = x, H = h) = E[exp(-s\,T)|W = x, X(0) = i, H = h]$$

$$= \begin{cases} exp(-sx/r_i), & \text{if } hr_i \geq x, \\ exp(-sh) \sum_{j=1}^n \dfrac{dK_{ij}(h)}{dK_i(h)}\,F_j^\sim(s), & \text{if } x > h\,r_i. \end{cases}$$

$$(7.17)$$

In (7.17), two mutually exclusive events are identified: if $hr_i \geq x$, then $T = x/r_i$ or if $hr_i < x$ then a transition occurs to state j $(j \neq i)$ and a different independent task with the same distribution is restarted. By unconditioning (7.17) with respect to W and H, we obtain:

$$F_i^\sim(s) = \int_{h=0}^\infty \int_{x=0}^{h/r_i} exp(-sx/r_i)\,dG(x)\,dK_i(h)$$

$$+ \int_{h=0}^\infty \int_{x=h/r_i}^\infty \sum_{j=1}^n F_j^\sim(s)\,exp(-sh)\,dG(x)\,dK_{ij}(h). \quad (7.18)$$

Solving the integrals in (7.18) completes the proof.   □

Corollary 7.2 *Under the assumptions of Theorem 7.2, given that $X(t)$ is a CTMC with infinitesimal generator $\mathbf{Q}$, the LST $F_i^\sim(s)$ satisfies the following equation (see [271]):*

$$F_i^\sim(s) = G^\sim\left(\frac{s + q_i}{r_i}\right) + \sum_{j=1;j\neq i}^n \frac{q_{ij}}{(s + q_i)}\left[1 - G^\sim\left(\frac{s + q_i}{r_i}\right)\right]F_j^\sim(s).$$

$$(7.19)$$

Theorem 7.3 *Given that $X(t)$ is a semi-Markov process and all the states are of pri type, the LST $F_i^\sim(s, x)$ satisfies the following equation:*

$$
\begin{aligned}
F_i^\sim(s, x) &= exp(-sx/r_i)\left[1 - K_i(x/r_i)\right] \\
&+ \sum_{j=1}^{n} F_j^\sim(s, x) \int_0^{x/r_i} exp(-sh)dK_{ij}(h). \quad (7.20)
\end{aligned}
$$

Proof. Conditioning on the sojourn time in the initial state $H = h$, let us define:

$$
F_i^\sim(s, x \mid H = h) = E[exp(-sT) \mid X = x, X(0) = i, H = h]
$$

$$
= \begin{cases}
exp(-sx/r_i), & \text{if } h\, r_i \geq x, \quad (7.21) \\
exp(-sh) \sum_{j=1}^{n} \dfrac{dK_{ij}(h)}{dK_i(h)} F_j^\sim(s, x), & \text{if } x > h\, r_i.
\end{cases}
$$

Unconditioning with respect to H yields equation (7.20). $\qquad \square$

Corollary 7.3 *Under the assumptions of Theorem 7.3, given that $X(t)$ is a CTMC with infinitesimal generator $\mathbf{Q}$, the LST $F_i^\sim(s, x)$ satisfies the following equation (see [271]):*

$$
\begin{aligned}
F_i^\sim(s, x) &= exp[-(s + q_i)\, x/r_i] \\
&+ \sum_{j=1; j \neq i}^{n} \frac{q_{ij}}{s + q_i}\left(1 - e^{-(s + q_i)x/r_i}\right) F_j^\sim(s, x). \quad (7.22)
\end{aligned}
$$

7.4 Applications and extensions

In this section we present some applications and extensions of the basic model, considered so far in the literature.

Binary reward variables

When the reward rates are constrained to be binary variables, a binary partition of the state space is induced. Classical reliability/availability models fall in this class. The conceptual framework, formulated in the previous sections, offers a unified view to subtle reliability problems in which the system catastrophic failure depends on the duration of the downtime. The problem has a long history in the reliability literature [139, 391, 381, 430] and can be formulated in terms of the completion time of a "virtual task" whose work requirement is equal to the assigned downtime threshold [357]. If the down state is of either *pri* or *prd* type, a fatal failure occurs as soon as a single downtime greater than the threshold is encountered, while if the down state is of *prs* type, the fatal failure occurs when the threshold level is exceeded by the total accumulated down time. Nicola, Bobbio and Trivedi [357] have calculated the completion time under more general conditions, and have derived several related measures from the knowledge of the completion time distribution.

State space partition in preemption classes

The expressions in Section 7.3 are derived under the simplifying hypothesis that all the structure states of $X(t)$ belong to the same preemption class. A natural and useful extension is to consider a partition of the state space into different preemption classes. The accumulation of the reward is thus resumed or reset according to the characteristics of the state just abandoned at the transition. [272] provides closed-form Laplace transform solutions when all three types of preemption policies are eventually present in the system, with a semi-Markov structure-state process. More general task execution processes can be modelled and estimated, and different kinds of failures can be taken into account. The same authors [356] have further extended their analysis, by considering a stream of jobs arriving at the server according to a Poisson process.

The completion time of programs

A specialized application of the above theory has been devoted to studying the execution time of programs on computing systems. In their pioneering work, Castillo and Siewiorek [65] have considered the time required to correctly execute a program, taking into account hardware reliability, software (operating system) reliability, the workload of the system while the program is executing, and the type and amount of resources required to execute the program. The hardware and software reliabilities and the workload and resource characteristics contribute to the definition of the random environment in which the task is performed, and are represented by $X(t)$.

The evaluation of the completion time of programs executed on degradable systems with different types of checkpointing mechanisms is the subject of [70]. By a combination of *prs* and *pri* kinds of interruptions, and the consideration of block structured programs, the authors were able to compare different recovery mechanisms at different levels of nesting in the program structure.

Multiple reward models

The simultaneous execution of parallel tasks with different work requirements on a computing system has been considered in [41, 44]. Each task α ($\alpha = 1, \ldots, \nu$) is served in each state i of $X(t)$ ($i = 1, \ldots, n$) at a different reward rate $r_{i\alpha}$. The reward rates are therefore grouped into a reward matrix, whose generic row $\mathbf{r}_i$ is the ν-dimensional vector representing how the total computational capacity of the system in state i is shared among the ν parallel tasks running in state i. On the other hand, the generic column $\mathbf{r}_\alpha$ contains the service rates at which task α is executed in the different structure-states in which the system operates. The minimal completion time has been derived in [44] under various combinations of preemption policies with $X(t)$ semi-Markovian.

7.5 Completion time and Petri nets

The functional $Y(t)$, which allowed us to define the completion time as the hitting time against an absorbing barrier (equation (7.2)), is a complex stochastic process even if the structure-state process $X(t)$ is a CTMC (Corollaries 7.1, 7.2 and 7.3). Stochastic Petri nets (SPNs) are usually restricted to be Markovian and therefore cannot be invoked to model and analyze the stochastic problem formulated in the previous sections. Recently, some attempts have been presented in the literature aimed at generalizing the concept of stochastic Petri nets by allowing the firing times to be generally distributed [4, 244, 78, 45]. The inclusion of non-exponential firing times poses intriguing problems about the interpretation of the evolution of the net versus time. A detailed discussion of the semantics of SPNs with generally distributed transition times can be found in [4]. We refer to this model as Generally Distributed Transitions SPN (GDT_SPN). The marking process underlying a GDT_SPN does not have, in general, a tractable analytical formulation. Therefore, various restrictions have been proposed in the literature. A particular case of non-Markovian SPN is the class of deterministic and stochastic Petri net (DSPN). DSPNs were introduced in [6] as extensions of generalized stochastic Petri nets (GSPNs) [5] where in each marking only a single enabled transition is allowed to have associated a deterministic firing time. DSPNs become of potential concern in the completion time analysis, when the structure-state process is a CTMC and the work requirement is a constant.

Several extensions of the original DSPN model have recently appeared in the literature [78, 45], aimed at including non-deterministic distributions in the model, and at accommodating more complex preemption policies for the general distributed transitions.

In the following subsections, we enumerate the features and the properties of the mentioned SPN-based models that are relevant in the context of the problems discussed in the present chapter, we provide a general framework for modelling completion time problems in terms of GDT_SPNs, and we conclude with a numerical example.

7.5.1   Generally Distributed Transitions SPN

Definition 7.1 *According to [4], a GDT_SPN is defined as a marked PN in which:*

1. *The set of transitions is partitioned into a subset of immediate transitions (thin bars) and a subset of timed transitions (thick bars). Immediate transitions fire in zero time and have priority over timed transitions [5].*

2. *To each timed transition t_k is assigned a generally distributed random firing time γ_k, with CDF $G_k(t)$, modelling the time needed to complete the activity associated to t_k.*

3. *An execution policy is defined, which specifies the way in which a transition is selected to fire (among those enabled in a given marking), and the way in which the GDT_SPN keeps track of the past history.*

The *execution policy* is needed to unequivocally determine a stochastic process associated to the PN. It comprises two specifications: a criterion to choose the next transition to fire (the *selection policy*), and a criterion to keep memory of the past history of the process (the *memory policy*). A natural choice to select the next transition to fire is according to a *race policy*: if more than one transition (of the same highest priority level) is enabled in a given marking, the transition fires whose associated random delay is statistically the minimum. The memory policy specifies how to recalculate the firing time distribution of a transition which has been disabled without firing when it is enabled again. In the exponential case, this problem is hidden by the memoryless property. Two alternative memory policies are considered:

- *Age memory*: An age variable a_k, associated with transition t_k, counts the time since the last firing epoch of t_k. When t_k is enabled, its firing distribution is calculated as the residual CDF of the associated random variable γ_k, conditioned to a_k.

- *Enabling memory*: The age variable a_k counts the time since the last epoch in which t_k has been enabled. When t_k is disabled (even without firing), a_k is reset.

Under the *age memory* policy the time spent in a PN transition accumulates whenever the transition is enabled and can be utilized to realize a *prs* preemption policy. Under the *enabling memory* policy the time spent in the transition is reset as soon as the transition is disabled and therefore can realize a *prd* preemption policy.

A numerically tractable realization of the GDT_SPN defined in Definition 7.1 is obtained by restricting the firing time random variables γ_k to be PH-distributed [45]. The non-Markovian process generated by the GDT_SPN is converted into a CTMC defined over an expanded state space. The cardinality of the expanded state space is of the order of the cross-product of the reachability set of the basic PN and the state spaces of the PH distributions of the γ_k random variables.

The program package ESP [100] realizes the GDT_SPN model with PH distributions. According to Definition 7.1, the program allows the user to assign a specific memory policy to each PN transition so that the different execution policies can be put to work. The important point about the ESP package is that the expanded CTMC is generated automatically from the model specifications. The generation of the expanded state space is driven by the different execution policies assigned by the user at the specification level.

The applicability of the GDT_SPN model with PH distributions to the completion time problem is legitimated by the following result proved in [48] and rederived by Neuts in [352]:

> *The class of PH distributions is closed with respect to the completion time problem in Markov reward models under any probabilistic mixing of* prs *and* prd *transitions.*

Hence, when the work requirement W is a PH random variable, or is approximated by a PH random variable, we are inside the area covered by the modelling power of the ESP package.

7.5.2   Deterministic SPN

Definition 7.2 *According to [6], a DSPN is defined as a marked PN in which:*

1. *The set of transitions is partitioned into a subset of immediate transitions, a subset of exponential transitions and a subset of deterministic transitions.*

2. *At most, a single deterministic transition is allowed to be enabled in each marking and the firing time of a deterministic transition is marking-independent.*

3. *The time elapsed in a deterministic transition cannot be remembered when the transition becomes disabled. The only allowed execution policy is the* race *policy with* enabling memory.

In [6], the steady-state probability distribution is the only addressed solution. An improved algorithm for the evaluation of the steady state probabilities has been successively proposed in [290, 291], and some structural extensions, with respect to the specifications of Definition 7.2, have been presented in [80]. However, the computation of the distribution of the completion time requires a transient analysis.

The DSPN model has been revisited in [74]: the stochastic process associated with the DSPN model is proved to be a Markov regenerative process and an analytical method for the derivation of both the transient and the steady-state solution is provided. The analytical solution is derived in the Laplace transform domain, whose inversion necessitates a numerical technique. The paper proposes the use of Jagerman's method [241], as adapted by Chimento and Trivedi [70].

An alternative numerical solution technique, based on the use of supplementary variables [97], was originally proposed in [164] for the steady state analysis and then extended in [159] to the transient analysis of particularly structured DSPNs.

7.5.3   Markov Regenerative SPN

A further extension, called the *Markov Regenerative SPN** (MRSPN*) model, has been developed in [75], where the structural restrictions implied in Definition 7.2 are retained, while replacing the deterministic transitions with generally distributed transitions. In particular, only the enabling memory policy can be assigned to the generally distributed transitions. This extension makes it possible to evaluate completion time problems in which the structure-state process is a CTMC, the work requirement is any random variable and the preemption policy is *prd*.

The supplementary variable approach to the same model has been discussed in [159] and a tool has been built based on this technique [162].

In order to relax the restriction on the enabling memory policy, Bobbio and Telek [46] have defined a new class of MRSPN based on the concept of non-overlapping dominant transitions. In this model, any two successive regeneration time points of the marking process correspond to the first enabling and to the firing (or disabling) of a single generally distributed transition called the dominant transition. The enabling cycles of the dominant transitions cannot overlap. This definition includes the possibility that the structure-state process is semi-Markov, and allows the accommodation of different preemption policies. The *prs* case has been introduced in [46, 456], and a specific algorithm for the steady-state analysis has been elaborated in [457]. Finally, the inclusion of the *pri* policy has been discussed in [38].

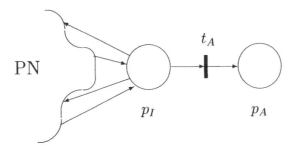

Figure 7.2: A completion time problem represented as a first marking time in place p_A

7.5.4 Modelling completion time by Petri nets

The completion time problem can be pictorially represented, at the SPN level, as a *first marking* problem. To this end, let us suppose that the reward rate is equal to one ($r = 1$) in all the states producing useful work. By this we physically mean that the task completes as soon as the total accumulated time spent in the markings producing useful work reaches the work requirement W. We introduce in the basic PN two additional places: an indicator place p_I and an absorbing place p_A (Figure 7.2).

The indicator place p_I is connected to the original SPN in such a way that it remains marked as long as the system is producing useful work. On the other hand, the absorbing place p_A is inserted to stop the execution of the net as soon as it becomes marked for the first time. The two places p_I and p_A are connected with each other by a single timed transition t_A. The random variable associated to the timed transition t_A coincides with the work requirement W.

Interpreting the model as a GDT_SPN, the transition firing occurs according to the semantics of the race policy: t_A fires when its associated firing time W is the minimum among the activities enabled in p_I. Hence, the epoch at which p_A becomes marked for the first time is the epoch at which the time elapsed in p_I exceeds W for the first time, and thus, by construction, is the completion time. In standard PN models, stopping the net usually requires additional elements, such as immediate transitions or inhibitor arcs. Using higher level nets, such as nets with enabling functions [83], the indicator, absorption and stopping property can be obtained by means of simpler and more natural specifications.

The semantics of the memory policies of a GDT_SPN [4] are suited to model different preemptive disciplines in the task completion time problem. This feature differentiates the GDT_SPN model from the DSPN where the enabling memory policy is the only available one. If transition t_A (Figure 7.2) follows a race policy with age memory, it fires as the total marking time accumulated in p_I exceeds W (independently of the number of times place p_I has become marked); from the point of view of the completion time problem, a *prs* policy is realized. If t_A follows a race policy with enabling memory, it fires the first time a continuous marked interval in place p_I (without interruptions) exceeds W. The GDT_SPN models a completion time problem with *prd* policy.

In the DSPN model, t_A is a deterministic transition. The semantics of the DSPN model enforce a preemptive repeat policy: each time place p_I is enabled again, a new task is started. Since in this case, the task requirement is deterministic, the *pri* and *prd* policies are coincident.

7.5.5 Numerical examples

We compare the results obtained from the GDT_SPN with PH distributions and the DSPN models, on two simple examples. The comparison is particularly significant to explore the flexibility and the accuracy of the PH approximation in a limiting case, since a deterministic variable is known to be typically non-PH.

Case 1: Bounded catastrophic breakdown

A system alternates between an up state (place p_1) and a down state (place p_2). Transition t_1 represents system failure (with failure rate λ) and transition t_2 system repair (with repair rate μ). A catastrophic (unsafe) condition is reached if and only if the time elapsed in the failed state exceeds a tolerance threshold W. This problem is represented in Figure 7.3, where place p_2 acts as indicator place and place p_A is the absorbing place representing the catastrophic condition.

Transition t_A is assigned the tolerance threshold W so that the catas-

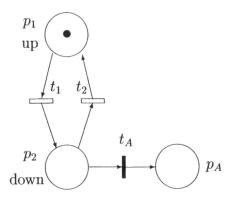

Figure 7.3: PN modelling the attainment of a catastrophic failure when the down time exceeds a critical threshold

trophic condition (token in p_A) is reached when the total down time exceeds W according to the assigned memory policy. If t_A is assigned an age memory policy, a *prs* strategy is realized since the time in p_A accumulates independently of the number of passages. On the other hand, if t_A is assigned an enabling memory policy, a *prd* strategy is realized. The distribution of the system lifetime can be interpreted as the distribution of the completion time of a "virtual task" of duration W, executed in place p_2. Since in the following we use the results obtained from [74] in the framework of DSPN models, only the *prd* policy can be considered.

Figure 7.4 shows the lifetime CDF with $\lambda = 0.001$ h^{-1}, $\mu = 0.1$ h^{-1} and with *prd* policy. The solid line represents the deterministic case with $W = 10$ h and is computed by numerically inverting the Laplace transform obtained from [74]. Dashed lines represent the cases in which W is Erlang with expected value $E[W] = 10$ h and increasing number of stages (2, 10 and 100, respectively). Since the resulting CDF is rather smooth also in the limiting deterministic case, the PH approximation becomes already close to the deterministic case with only a small number of stages.

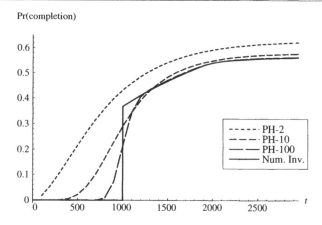

Figure 7.4: CDF of the lifetime of a two-state system subject to a bounded catastrophic breakdown

Case 2: Completion time with bounded catastrophic breakdown

A system that can reach a catastrophic condition, as in the previous case, executes a task of length Z. The PN modelling the system is shown in Figure 7.5 [357], where t_B is assigned a firing time equal to the work requirement Z, and a token in p_B stops the net as soon as the task execution is completed. The PN models a competing multiple completion time example, and the analysis is aimed at evaluating the defective CDF of completing the task before reaching the catastrophic state (a token arrives in p_B before one arrives in p_A). Computations are performed supposing that the up state is prd and the down state is prs. The deterministic case is solved by numerically inverting the closed-form Laplace transform given in [357].

Figure 7.6 compares the case in which both barrier levels W and Z are deterministic with the case in which both are Erlang of the same increasing order (2, 10 and 100 stages, respectively). Failure and repair rates are as in Case 1: $E[W] = 10$ h, and $E[Z] = 1000$ h. As can be observed, abrupt changes in the CDF shape require PH variables of very high order.

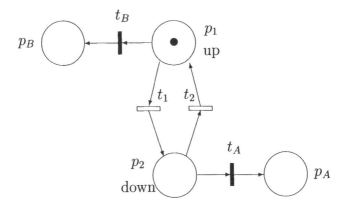

Figure 7.5: PN modelling the completion time of a task on a two-state server subject to a bounded catastrophic breakdown

7.6   Conclusion

The distribution of the task completion time is a performability measure that characterizes the quality of the service in a dependable computing system.

Interpreting the task completion time as the hitting time of a suitable cumulative functional against an absorbing barrier provides a flexible and useful representation of the problem in many applications. In fact, various kinds of policies can be accommodated for modelling the interruption and the subsequent recovery of the execution of a task. Three different preemption policies have been extensively discussed and closed-form expressions for the CDF of the completion time have been derived in the Laplace transform domain.

Non-exponential stochastic Petri nets can provide a graphical descriptive tool for the completion time analysis. In particular, recently proposed SPN models are reviewed. These models allow the inclusion to some extent of generally distributed firing times. A numerical example compares the results obtained from two specific models. In the first model, the firing time distributions are allowed to be of PH type, while in the second model a single transition in each marking is allowed to have a deterministic firing time, with all the other firing times being exponential.

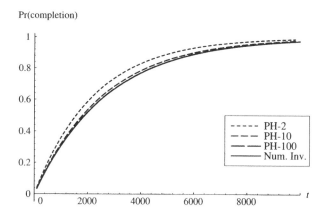

Figure 7.6: CDF of the completion time of a two-state server with *prd* up state and *prs* down state

Acknowledgements

The work of Andrea Bobbio was partially supported by CNR grant no. 96.01939.CT12. Miklós Telek was partially supported by OTKA grant no. T-16637.

Chapter 8

Rare Event Simulation

Perwez Shahabuddin

We review simulation techniques for estimating dependability and performance measures in stochastic models of computer and communication systems. Naive simulation is not efficient in this setting as most events of interest (for example, a system failure event) are rare. Hence special simulation techniques need to be used to speed up the occurrence of these rare events. We describe and illustrate two classes of techniques: importance sampling and importance splitting (the latter is also called RESTART or multilevel splitting). The first one simulates the system using a new probability dynamics under which the event of interest happens more frequently; the estimate is then adjusted to make it unbiased. Coming up with the new probability dynamics is the main challenge here; any arbitrary probability dynamics that make the event of interest happen more frequently may even lead to much worse estimation compared to naive simulation. The other class of techniques is based on the concept of splitting the simulated sample paths of the stochastic process when they "approach" the set of rare events; this way more samples of the rare event are obtained. When to split and how many splits are questions of interest

here. We illustrate the application of these techniques to some
common reliability and queueing network models of computer
and communication systems.

8.1   Introduction

IN this chapter we review simulation techniques used for estimating dependability measures in stochastic models of computer and communication systems. The common theme that runs throughout this chapter is
that for systems that are highly dependable, the events of interest are rare,
and so naive simulation is highly inefficient. Hence, special techniques have
to be used to speed up the occurrence of these rare events.

This chapter is by no means a comprehensive survey of all rare event
simulation techniques, nor does it present a complete reference list of all
the contributions in this area. However, an attempt has been made to give
some of the basic concepts and algorithms used for different types of computer and communication system models. For those types of models for
which this has not been possible due to space constraints, an attempt has
been made to point to the latest references, so that an interested reader
may follow up. The reader is referred to Nakayama [347], Heidelberger
[218] and Nicola, Shahabuddin and Nakayama [364] for more comprehensive surveys of fast simulation techniques applied to reliability models, and
to Heidelberger [218] and Asmussen and Rubinstein [14] for techniques applied to queueing and queueing network models.

Estimations of the small probabilities of rare events are required in the
design of many computer and communication systems. Consider the case
of a telecommunication network. It is customary to model such systems
as networks of queues, with each queue having a buffer of finite capacity.
Data packets that arrive at a queue when its buffer is full are lost. The
rare event of interest may be the event of a packet being lost. Current
regulations stipulate that the probability of packet loss should not exceed
10^{-9}. While packet loss may corrupt data during transmission, large delays
in transmission may affect the quality of certain real-time applications, e.g.,
video-on-demand or video conferencing. Hence it may also be of interest to
estimate the probability of large delays in packet transmission. Since these
probabilities are required to be small, estimating such probabilities is also

a rare event simulation problem. Alternatively, in a reliability model of a spacecraft computer, we may be interested in estimating the probability of the event that the system fails before mission completion. Naturally, one would want this probability to be extremely low. The main problem with using standard simulation to estimate such small probabilities is that a large number of events have to be simulated in the model before any samples of the rare event may occur. Hence special simulation techniques are needed to make the events of interest occur more frequently.

Importance sampling is a technique that can be used for this purpose. This technique was initially used in the area of Monte Carlo integration (see, e.g., Kahn and Marshall [252]). An extension of the basic concept to stochastic processes may be found, for example, in Glynn and Iglehart [177]. In importance sampling, the stochastic model is simulated with a new probability dynamics that make the events of interest occur more frequently. The sample value is then adjusted to make the final estimate unbiased. However, choosing any change of measure that makes the event of interest occur frequently is not enough; *how* it is made to happen more frequently is also very important. For example, an arbitrary change of measure that makes the rare event happen more frequently may give an estimator with an infinite variance (see, e.g., Juneja and Shahabuddin [248]). Thus the main problem in importance sampling is to come up with an appropriate change of measure for the rare event simulation problem at hand. Different classes of stochastic models may use changes of measure that are totally different in nature.

Another method which makes rare events happen more frequently is a technique introduced in Bayes [26] which is closely related to methods described by Kahn and Harris [251]. In standard simulation, the stochastic process being simulated wastes a lot of time in a region of the state space which is "far away" from the set of rare events, i.e, from where the chance of it entering the rare set is extremely low. In Bayes's approach, a region of the state space that is "closer" to the rare set is defined. Each time the process reaches this region, from the "faraway" region, many identical copies of this process are generated. In simulation terminology this is called *splitting*. Each of the split copies is simulated until it exits back into the "faraway" region. From there on, only one of the split copies is continued until another entrance into the "closer" region. This way we get more instances of the stochastic process spending time in a region where the rare event is more likely to occur.

There are a few software-based modelling tools which use these rare event simulation techniques. SAVE (see Blum *et al.* [36]) incorporates a provably efficient importance sampling heuristic for reliability models called balanced failure biasing. UltraSAN (see Obal and Sanders [366]) gives the user the capability to specify an importance sampling change of measure for certain classes of stochastic models. Importance sampling methods for estimating the normalization constants of multiclass closed queueing networks are incorporated in MonteQueue (see Ross, Tsang and Wang [392]). A version of the splitting method has been implemented in ASTRO (see Villen-Altamirano and Villen-Altamirano [468]).

The rest of the chapter is organized as follows. The problem statement and the quantities to be estimated in their most abstract form (i.e., without any reference to particular models) are given in Section 8.2. In Section 8.3 we illustrate the rare event simulation problem. In Section 8.4, the general concept of importance sampling is described. Importance sampling for reliability models is presented in Section 8.5.1, and for queueing models in Section 8.5.2. Applications of importance sampling for other types of stochastic models have been summarized in Section 8.5.3. Section 8.6 finally describes the splitting technique.

8.2   Problem statement

Consider a stochastic process $\{X(s) : s \geq 0\}$ on state space $\mathcal{S}$. We partition $\mathcal{S}$ into two subsets: $\mathcal{S} = \mathcal{G} \cup \mathcal{B}$ where $\mathcal{B}$ is the set of system states which are rare and of interest, and $\mathcal{G} = \mathcal{S}/\mathcal{B} \equiv \bar{\mathcal{B}}$. Suppose the process reaches steady state, i.e. $X(s) \Rightarrow X_\infty$ as $s \to \infty$, for some random variable X_∞. One measure of interest is estimating $\alpha = E(f(X_\infty))$ where $f(x) = 1_{\{x \in \mathcal{B}\}}$. From the physical point of view, this is the long-run fraction of time the process spends in the rare state $\mathcal{B}$. Sometimes we may also be interested in the mean time between visits to the set $\mathcal{B}$, while the process is in steady state. We denote this by β.

We may also wish to estimate certain transient quantities such as the fraction of time during $[0, t]$ the process spends in the set $\mathcal{B}$, i.e.,

$$\alpha(t) = E(\int_{s=0}^{t} f(X(s))ds/t).$$

Let τ_B be the first time the process hits the set B. Then $E(\tau_B)$ and $P(\tau_B < t)$ are also measures of interest in many situations.

8.3 Rare event simulation

Here we illustrate mathematically the basic problem of rare event simulation. Let Z be a random entity with probability measure $p(\cdot)$ on its sample space Ω and let $\mathcal{R}$ be a rare (under $p(\cdot)$) subset of the sample space. The problem may be to estimate $\gamma = P(\mathcal{R}) \equiv E_p(1_{\{Z \in \mathcal{R}\}})$ where the subscript in the expectation denotes the probability measure assigned to the random variable Z. Systems which have rare events are characterized by a rarity parameter ϵ so that as $\epsilon \to 0$, $\gamma \to 0$. For example, in a reliable system with highly reliable components, ϵ may be the maximum failure rate of components in the system. For a queueing system with buffer size B, we can set $\epsilon = 1/B$ so that as $\epsilon \to 0$ the buffer overflow event becomes more rare.

In standard simulation, we generate n samples of the random variable Z, say $Z_1, Z_2, \ldots Z_n$ and estimate γ by using $\hat{\gamma} \equiv \sum_{i=1}^{n} 1_{\{Z_i \in \mathcal{R}\}}/n$. For fixed n, the half width (HW_p) of the confidence interval is (approximately) directly proportional to $\sqrt{\mathrm{var}_p(1_{\{Z \in \mathcal{R}\}})} = \sqrt{\gamma - \gamma^2} \approx \sqrt{\gamma}$ for small γ. Consequently, the relative error $RE_p \equiv HW_p/\gamma$ is directly proportional to $\sqrt{\mathrm{var}_p(1_{\{Z \in \mathcal{R}\}})}/\gamma$. It is then easy to see that $RE_p \to \infty$ as $\epsilon \to 0$. Equivalently, the simulation run length n required to achieve a fixed relative error RE_p goes to ∞ as $\epsilon \to 0$.

A related problem is the estimation of $\psi = E_p(W(Z)1_{\{Z \in \mathcal{R}\}})$ where $W(\cdot)$ is some function with domain Ω. Since $\mathcal{R}$ is rare, this expectation tends to be small and difficult to estimate. From the representation of ψ given above, it will not seem surprising that fast simulation techniques that work for the estimation of γ also seem to work for ψ.

8.4 Importance sampling

Let $p'(\cdot)$ be another probability measure on the sample space of Z, so that $p'(z) > 0$ whenever $p(z) > 0$ for all x in $\mathcal{R}$. Then

$$\gamma = \int_{z \in \Omega} 1_{\{z \in \mathcal{R}\}} p(z) dz = E_{p'}(1_{\{Z \in \mathcal{R}\}} L_{p'}(Z)) \tag{8.1}$$

where the subscript in the expectation denotes the probability measure assigned to Z and $L_{p'}(\cdot)$ is the likelihood ratio, i.e., $L_{p'}(z) \equiv p(z)/p'(z)$ whenever $p'(z) > 0$ and 0 otherwise. Equation (8.1) suggests that we can use $p'(\cdot)$ instead of $p(\cdot)$ to generate n samples of Z and then use $\hat{\gamma} \equiv \sum_{i=1}^{n} 1_{\{Z_i \in \mathcal{R}\}} L_{p'}(Z_i)/n$ as an unbiased estimator of γ. This is called importance sampling. The problem is to choose a $p'(\cdot)$ so that

$$\mathrm{var}_{p'}(1_{\{Z \in \mathcal{R}\}} L_{p'}(Z)) \ll \mathrm{var}_{p}(1_{\{Z \in \mathcal{R}\}}).$$

In many rare event simulations with importance sampling, an attempt is made to come up with changes of measure $p'(\cdot)$, so that the relative error, $RE_{p'}$, remains bounded as $\epsilon \to 0$. This is known as the *bounded relative error* (BRE) property.

Note that $E_{P'}(1_{\{Z \in \mathcal{R}\}} L_{p'}^2(Z)) \geq \gamma^2$. Hence another condition is to require that

$$\limsup_{\epsilon \to 0} \frac{\log[E_{P'}(1_{\{Z \in \mathcal{R}\}} L_{p'}^2(Z))]}{\log[\gamma]} \geq 2,$$

i.e., the exponential rate of decay of $E_{P'}(1_{\{Z \in \mathcal{R}\}} L_{p'}^2(Z))$ with decreasing ϵ is the slowest possible, thus guaranteeing good variance reduction over naive simulation. This is termed asympotic optimality and it can be shown to be weaker than the BRE property mentioned above.

To apply importance sampling to estimate the measures given in Section 8.2, one first has to represent these measure in terms of small probabilities and expectations of the type given in Section 8.3. First consider the case where $\{X(s) : s \geq 0\}$ is regenerative [98]. Pick a regenerative state that is frequently visited and let τ be the corresponding regenerative cycle time. Define W to be the amount of time in a regenerative cycle that the process spends in the rare set $\mathcal{B}$. Then α may be represented as the ratio $E(W)/E(\tau)$. In most cases $E(\tau)$ is not small and is thus easy to estimate using standard simulation. However, most samples of W are zero and thus we need importance sampling to estimate $E(W)$ accurately. In relation to Section 8.3, the Z corresponds to a random sample path *in a regenerative cycle* of the above stochastic process, the $p(\cdot)$ is the original probability measure on these sample paths, $\mathcal{R}$ is the set of these sample paths that visit $\mathcal{B}$, and $W(Z) \equiv W$ is the amount of time the sample path Z spends in $\mathcal{B}$. Thus the rare event simulation problem becomes a problem of estimating $E_p(W(Z)1_{\{Z \in \mathcal{R}\}}) = E_p(W(Z)) \equiv E(W)$. The changes of measure

$p'(\cdot)$ used are such that they induce a drift in the stochastic process towards $\mathcal{B}$ so that the chance of the rare event $\mathcal{R}$ happening is increased. However, once $\mathcal{B}$ is visited, the stochastic process is simulated with the usual dynamics so that the regenerative cycle soon completes (Goyal *et al.* [185]).

Given that at time $t = 0$ the system is in the regenerative state, the expectation $E(\tau_B)$ defined in Section 8.2 may be represented as the ratio $E(\tau_{min})/P(\mathcal{R})$ (see, e.g., Keilson [257]). Here $P(\mathcal{R})$ is the probability of the rare event $\mathcal{R}$, i.e., of hitting $\mathcal{B}$ in a regenerative cycle, and τ_{min} is the time to hit either $\mathcal{B}$ or the regenerative state given that the process starts in the regenerative state. Again, $E(\tau_{min})$ is easy to estimate using standard simulation. However, we have to use importance sampling to estimate $P(\mathcal{R})$.

The other transient measures are naturally defined in terms of small probabilities and expectations, if the time horizon is small. However, for large time horizons, the importance sampling variance may grow exponentially with time (see Glynn [175]). In those cases, some other representations of the transient measures may prove useful (see, e.g., Shahabuddin and Nakayama [427]).

Now consider the case where $\{X(s) : s \geq 0\}$ is non-regenerative (this also applies when the regenerative cycles are very long, so that the regenerative simulation procedure cannot be used effectively). One can still use a ratio representation of the steady-state measure. Let $\mathcal{A}$ be a state or a set of states that are visited quite frequently in the simulation. Define an $\mathcal{A}$-cycle to start whenever the process enters the set $\mathcal{A}$. Then the ratio formula $\alpha = E(W)/E(Z)$ still holds, where now $W \equiv W(Z)$ is the amount of time the process spends in the set $\mathcal{B}$ in an $\mathcal{A}$-cycle and $\tau = \tau(Z)$ represents the duration of an $\mathcal{A}$-cycle, given that the process is in steady state (e.g., Cogburn [90]). The actual simulation procedure uses a splitting technique combined with batch means [361]. We first run a few $\mathcal{A}$-cycles so that the system (approximately) reaches steady state. After that, each time an $\mathcal{A}$-cycle starts, we split a process from the original process. The split process uses the change of measure as prescribed by the importance sampling. We use this split $\mathcal{A}$-cycle to get a sample of $W(Z)$ and $L_{p'}(Z)$ and use the original $\mathcal{A}$-cycle to get a sample of $\tau(Z)$. Since the successive $\mathcal{A}$-cycles are generally not independent, we have to use the procedure of batch means to build confidence intervals. A similar splitting idea was used in Al-Qaq,

Devetsikiotis and Townsend [11] for estimating bit error rates over certain communication channels.

We can also use a ratio representation to estimate β which was defined in Section 8.2: $\beta = E(\tau)/E(N)$ where N is the number of visits to $\mathcal{B}$ during an $\mathcal{A}$-cycle (Glynn *et al.* [176]). Again, $E(N)$ is the small expectation which we have to estimate using importance sampling.

8.5   Applications of importance sampling

8.5.1   Reliability models

Consider the fairly general class of reliability models considered in Blum *et al.* [36]. These are systems consisting of components that fail and get repaired. Components are not independent in the sense that they share repair units, they have operational/repair dependencies and there may be failure propagation (i.e., the failure of a component may cause another component to fail instantaneously). If we assume that component failure times and repair times are exponentially distributed, then the system can be modelled as a continuous time Markov chain (CTMC) $\{X(s) : s \geq 0\}$. For example, in the simplest such system, $X(s) = (X_1(s), X_2(s), \ldots, X_N(s))$ where $X_i(s)$ may be considered to be the number of components of type i that are up and N is the total number of component types. A transition of the CTMC corresponds to either a component failure transition or a component repair transition. In mathematical models of highly reliable systems, the failure rate of a component, say component i, is represented as $\lambda_i \epsilon^{r_i}$ where ϵ is the rarity parameter, and r_i, λ_i are positive constants, i.e., independent of ϵ. The r_i's may be different if the system is "unbalanced", i.e., components have failure rates that are of different orders of magnitude. Since the repair rates are comparatively large, they are represented by a constant.

First consider steady state estimation. For this purpose one can simulate the embedded discrete time Markov chain of the CTMC. Let $\mathbf{P} = \{P_{x,y} : x, y \in \mathcal{S}\}$ denote the transition matrix for this Markov chain. The regenerative state is taken to be the one in which all components are up.

In relation to Section 8.2, the set $\mathcal{B}$ represents the set of states of the CTMC in which the system is considered failed or unavailable. The α defined in Section 8.2 is thus the steady state unavailability, β is the steady state mean time between system failures, and $E(\tau_B)$ is the mean time to system failure from the "all-components-functioning" state. The changes of measure used here are called "failure biasing heuristics" and correspond to simulating the system using a new probability transition matrix $\mathbf{P}' = \{P'_{x,y} : x, y \in \mathcal{S}\}$, with the property that for any states x, y, $P'_{x,y} > 0$ if $P_{x,y} > 0$. If a state in $\mathcal{B}$ is visited before the regenerative cycle completes, then the transition matrix $\mathbf{P}$ is used for the remainder of the cycle. The intuitive idea behind these heuristics is to artificially make failure transitions happen much more frequently than in the actual system. The original heuristic was introduced in [284] and is now called simple failure biasing in the literature. However, this heuristic does not have the BRE property for unbalanced systems [425]. A modified technique called balanced failure biasing [185, 425, 424] has been proven to have the BRE property in [425, 424]. The following compact representation has been taken partly from [347].

Algorithm 8.1: Balanced failure biasing

- From any state x, let $\Lambda_F(x)$ (*cf.* $\Lambda_R(x)$) be the set of transitions (x, y) that correspond to component failure (*cf.* repair) transitions. Let $p_F(x) \equiv \sum_{y \in \Lambda_F(x)} P_{x,y}$ and $p_R(x) \equiv \sum_{y \in \Lambda_R(x)} P_{x,y}$. Define $I_{x,y} = 1$ if $P_{x,y} > 0$ and $I_{x,y} = 0$ otherwise. For any state x, define $n_F(x)$ to be the number of failure transitions possible (under $\mathbf{P}$) from x. Let p^*, $0 < p^* < 1$, be a constant. In practice, $0.5 \le p^* \le 0.9$.

- If $p_R(x) > 0$ then

$$P'_{xy} = \begin{cases} p_* I_{x,y}/n_F(x), & \text{if } (x, y) \in \Lambda_F(x), \\ (1 - p_*)P_{x,y}/p_R(x), & \text{if } (x, y) \in \Lambda_R(x), \\ 0, & \text{otherwise.} \end{cases}$$

- If $p_R(x) = 0$, let $P'_{x,y} = I_{x,y}/n_F(x)$, if $(x, y) \in \Lambda_F(x)$ and $P_{x,y} = 0$ otherwise.

A crucial assumption used in Shahabuddin [424] to prove the BRE property of balanced failure biasing is that all states of the Markov chain, except the state in which all components are up, have at least one component repair transition. However, in many practical situations this may not be the case. For example, often when a component fails, we may want to defer the repair of that component to a time when some other components have also failed. It may be more cost effective to call upon a repair unit at that time who will repair all the failed components in one visit. In many cases, we have field replaceable units (FRUs) comprising two or more components, e.g. a processor card with multiple processors. In practice we wait until a certain number of components of the FRU fail before changing the whole FRU (this is an example of "group repair"). In situations where this assumption does not hold, balanced failure biasing may give infinite variance (shown in [246]). Juneja and Shahabuddin [247, 249] developed improved failure biasing schemes for the fast simulation of such systems.

Another failure biasing heuristic based on the concept of failure distances may be found in Carrasco [63]. A detailed investigation of the conditions on systems under which failure biasing heuristics give bounded relative error was made by Nakayama [346, 345]. However, so far it appears difficult to use these results in practice. Some additional results in this regard may be found in Strickland [447]. For results and references on derivative estimation the reader is referred to Nakayama [344].

Now we describe the estimation of transient dependability measures. In the context of the reliability model described in this section, the $\alpha(t)$ described in Section 8.2 corresponds to the expected interval unavailability and $P(\tau_B < t)$ corresponds to the unreliability. For estimating these transient measures, just using failure biasing is not enough. We also have to use some mechanism to ensure that the first transition happens before time t. This is termed forcing and was introduced in Lewis and Bohm [284]. It is shown in [425, 427, 428] that forcing combined with failure biasing produces BRE in the estimation of the reliability and the interval availability for cases where t is small (t is either of the same order as the regenerative cycle time or smaller). However, for cases where t is large the relative error tends to infinity. For such cases, a method based on estimating Laplace transform functions of the transient measure is studied in Carrasco [63] and another approach based on estimating bounds to the transient measure (rather than estimating the actual measure) is studied in Shahabuddin and Nakayama [425, 427, 428]; the latter also deals with the estimation of derivatives of transient measures.

For non-Markovian models, an importance sampling approach based on rescheduling failure events is given in Nicola *et al.* [359]. Two other approaches for estimating the unreliability, one based on uniformization and the other called exponential transformation, were introduced in [360]. In [221], these schemes were shown to have the BRE property under fairly general conditions. Using the $\mathcal{A}$-cycle method described at the end of Section 8.4, these approaches were extended to cover estimation of the steady state unavailability and the steady state mean time between failures (MTBF) in [361] and [176], respectively.

Some work has also been done in the area of estimation of unreliability in a network with independent components. A network is modelled as a graph whose edges represent the components. The network is said to fail if the connectivity between two (disjoint) sets of nodes is lost. Again, if the components are highly reliable, then the chance of a network failure is very small; detailed importance sampling schemes can be found in [142, 288, 394], among many others.

There are two ways in which a system composed of components can be made highly dependable in a cost-effective manner. The first is to use components that are highly reliable, and have low inbuilt redundancies in the system. The second is to build in significant redundancies but use cheap components that are not highly reliable. In Shultes [435] and Alexopoulos and Shultes [7, 8] it has been experimentally shown that the methods described above may not work well for systems of the second type. These authors introduce the "balanced likelihood ratio method" for effectively simulating steady state dependability measures in Markovian models of such systems. This attempts to cancel terms of the likelihood ratio within a regenerative cycle by defining the importance sampling probabilities for events in such a way that the contribution to the likelihood ratio from a repair event cancels those from a failure event that occurred previously in the same regenerative cycle.

8.5.2 Queueing models

Queues with identically distributed renewal inputs

Changes of measures for queueing models that have the asymptotic optimality property (for most of these models, it is very difficult to come up with techniques for which one can prove the BRE property) have been proposed and studied in Cottrell, Fort and Malgouyres [93], Parekh and Walrand [369] and Sadowsky [398], among many others. Most of these provably efficient changes of measure apply to a single server queueing system where the arrival stream constitutes an independent, identically distributed (i.i.d.) renewal process. The measures of interest have been the tail distribution of the steady state waiting time and queue length in systems with infinite buffer, and the steady-state customer loss probability in systems with finite buffer. The change of measure is based on "exponentially tilting" the arrival and the service distribution. Let $F_A(\cdot)$ (cf. $F_S(\cdot)$) denote the original interarrival time (cf. service time) distribution and let $M_A(\cdot)$ (cf. $M_S(\cdot)$) be its moment generating function. The new interarrival time (cf. service time) distribution $\tilde{F}_A(\cdot)$ (cf. $\tilde{F}_S(\cdot)$) corresponding to the provably efficient change of measure is determined as follows.

Algorithm 8.2: GI/GI/1 queue

- Let θ^* be the solution of

$$M_A(-\theta)M_S(\theta) = 1, \ \theta > 0.$$

- Then

$$d\tilde{F}_A(x) = e^{-\theta^* x} dF_A(x)/M_A(\theta^*)$$

   and

$$d\tilde{F}_S(x) = e^{\theta^* x} dF_S(x)/M_S(\theta^*).$$

For the M|M|1 queueing system with arrival rate λ and service rate μ (with $\lambda/\mu < 1$) this change of measure corresponds to interchanging the arrival rate and the service rate. Note that this makes the queue unstable so that large queue lengths are reached much faster. Sadowsky [398] presents a provably efficient change of measure for the GI|GI|m queueing system.

Extensions of these provably efficient changes of measure to networks have been few and apply mainly to Markovian tandem networks (e.g., Glasserman and Kou [172]). Heuristical approaches for fast simulation of more general networks, based on a large deviations approach, may be found in Parekh and Walrand [369] and Frater, Lennon and Anderson [148]. Ross, Tsang and Wang [392] have used importance sampling to estimate the normalizing constant which occurs in the solution of multiclass product-form closed queueing networks.

Queues with correlated arrival processes

Provably efficient changes of measure for discrete time queues with auto-correlated arrival processes were studied in Chang *et al.* [68] and continuous time versions in Juneja [245] (see, e.g., Lehtonen and Nyrhinen [278] for analogous concepts in the context of risk analysis). Fast simulation of Markov fluid models of such queues has been studied in Kesidis and Walrand [263] and Mandjes and Ridder [299]. Chang *et al.* [68] also linked fast simulation techniques for ATM switches to the concept of effective bandwidth of the arrival sources, thus generalizing the class of source models that can be handled and allowing the study to be extended to the class of intree networks. Some critical concepts in [68], dealing with effective bandwidths in intree networks, were also developed independently in [115]. Further experimentation with larger models using the simulation technique of [68, 67] has been done in l'Ecuyer and Champoux [275] and extensions to the basic simulation technique have been been made in Falkner, Devetsikiotis and Lambadaris [140]. Since the literature on this subject is vast, we just present the algorithm for a simple discrete-time queue that is fed by a Markov modulated arrival process (MMAP).

Consider a discrete time queue that is fed by K external sources, each of which is an MMAP. For simplicity, we consider the simplest form of such an arrival process. Let the kth source be in any of the M_k states $\{0, 1, \ldots M_{k-1}\}$. Let $Y_k(t)$ be the state of the kth source after time t, and let $p_k(i, j) = P(Y_k(t+1) = j | Y_k(t) = i)$. Let the number of packets a source transmits per unit of discrete time, $a_k(t)$, be equal to the current state of the source and let $a(t) = \sum_{k=1}^{K} a_k(t)$ be the total arrival to the queue in that unit of discrete time. We assume that the queue has the capacity to dispatch c packets every unit of discrete time. Let B denote the size of the buffer. Then the number of packets in the system at time t is governed by the following Lindley type recursion: $Q(t+1) = (\min(Q(t) + a(t+1), B) - c)^+$. The problem is to estimate the steady state probability of packet loss when $\epsilon \equiv 1/B$ is small. In this case we let the set $\mathcal{A}$ (corresponding to an $\mathcal{A}$-cycle) be the set of states of the extended Markov chain $(Q(t), Y_1(t) \ldots Y_K(t))$ that have $Q(t) = 0$. Let $\lambda_{k,\theta}$ be the spectral radius of the matrix that has elements $\mathcal{A}_k(i, j) = e^{\theta j} p_k(i, j)$ and let $h_{k,\theta}(j)$ be the corresponding eigenvector. The provably efficient change of measure corresponds to doing a sort of exponential tilting to the MMAPs (the service rate c remains unchanged). The new transition matrix for the kth MMAP, $p'_k(i,j)$, can be determined as follows.

Algorithm 8.3: Queues with MMAP arrivals

- Let θ^* be the solution of the equation $\sum_{k=1}^{K} log(\lambda_{k,\theta}) = c$ and $\theta > 0$.

- Then $p'_k(i,j) = e^{\theta j} p_k(i,j) h_{k,\theta}(j)/\lambda_{k,\theta} h_{k,\theta}(i)$.

8.5.3 Other models

Importance sampling has also been used in the estimation of the bit-error rate in digital communication systems. The reader is referred to Al-Qaq, Devetsikiotis and Townsend [11] for a list of references in this area. For importance sampling applied to general Markov chains, refer to Andradottir, Heyman and Ott [13], Glynn [175] and references therein.

8.6 The splitting method

This splitting method was introduced by Bayes [26], who referred to it as "importance sampling", as it does require sampling from a region of importance. However, in the current literature, the definition of importance sampling no longer seems to include this method. Hence we think that a more appropriate term for it may be "importance splitting".

Consider the stochastic process $\{X(s) : s \geq 0\}$ mentioned in Section 8.2, where the problem is to estimate α. The usual method is to first simulate the process until it (approximately) reaches steady state. After that, we simulate it for an interval of time t. For convenience, assume that at $s = 0$ the process is in steady state. Then $Y = (\int_{s=0}^{t} f(X(s))ds)/t$ gives an unbiased (one sample) estimate of α. One can use either the replication–deletion method or the batch means method to construct confidence intervals.

Let $\mathcal{C} \subset \mathcal{S}$ be such that $\mathcal{B} \subset \mathcal{C}$, and the steady state probability of being in states in $\mathcal{C}$ is not as small as α. By an "upcrossing" we will mean the stochastic process going from $\bar{\mathcal{C}}(\equiv \mathcal{S}/\mathcal{C})$ to $\mathcal{C}$. A "downcrossing" will mean the opposite. The following is a polished version of the algorithm in Bayes [26].

Algorithm 8.4: Splitting method

1. Set $j = 0$. Set simulation time $s = 0$, and the cumulator $sum = 0$.

2. Simulate one copy of the process until the next upcrossing. Update $j \leftarrow j + 1$ and set s_j to be the absolute time of this upcrossing. If $s_j < t$, update $s \leftarrow s_j$; otherwise end simulation and go to Step 5.

3. At s_j generate R split processes, each with the starting state $X(s_j)$, and simulate each split process until a downcrossing. Let Δ_r be the amount of this elapsed time (after s_j) for the rth split process and let Y_r be the amount of time in the interval $[s_j, \min\{s_j + \Delta_r, t\}]$ that the rth split process spends in the set $\mathcal{B}$. Let $\bar{\Delta} = \sum_{r=1}^{R} \Delta_r / R$. Update $sum \leftarrow sum + \sum_{r=1}^{R} Y_r / R$. Advance the simulation time to $s \leftarrow s + \bar{\Delta}$.

4. If $s > t$ then end the simulation and go to Step 5; otherwise set $X(s)$ equal to the state at the downcrossing of the Rth split path. Go to Step 2.

5. Let $\tilde{Y}$ be the amount of time in the interval $[0, \min\{s_1, t\}]$ that the process spent in $\mathcal{B}$. Form the estimator $\hat{\alpha} = (\tilde{Y} + sum)/t$.

Bayes [26] called the boundary between $\mathcal{C}$ and $\bar{\mathcal{C}}$ the "importance level". A possible generalization of this scheme to the case of multiple importance levels was also mentioned.

Hopmans and Kleijnen [232] investigated the above algorithm in detail (for the one-dimensional, one-level case) using a regenerative assumption, i.e., the system regenerates each time we have an upcrossing (or a downcrossing). In that sense the algorithm they use is slightly different in its execution from what is given above. By conducting a variance analysis they determined the optimum R. They applied it to a complex telecommunication system model but they were not very satisfied with the improvement in efficiency obtained. Villen-Altamirano and Villen-Altamirano [467] revisited this idea and proposed modified schemes to the one in Bayes [26] which they called RESTART. One main difference from Bayes's algorithm is that in Step 3, instead of s being updated to $s + \bar{\Delta}$, it is updated to $s + \Delta_R$. They also did an approximate variance analysis of their scheme to determine the optimum R and the optimum placement of the level, and then computed the efficiency gain obtained. Experiments using this scheme produced significant variance reduction on particular examples. Generalization of the scheme and the variance analysis to the multi-level case was done in Villen-Altamirano *et al.* [469] and adaptations to transient estimation were done in [468].

Recently, Glasserman *et al.* [171] described another version of the splitting technique for the multilevel case, and gave rigorous conditions under which the technique is asymptotically optimal for rare event simulation. One of the main restrictions imposed on the model is that there may only be a finite number of ways of achieving each level, and this generally restricts the treatment to models with infinite state spaces in only one dimension. The results suggest that the proper implementation of splitting in higher dimensions requires an understanding of the way rare events occur, i.e., the large deviation behaviour, not unlike what is often needed to use importance sampling.

8.7   Acknowledgements

This work was supported in part by NSF Career Award Grant DMI 9625297. A preliminary version of this work appeared in Shahabuddin [426] and was done while the author was at the IBM T.J. Watson Research Center, Yorktown Heights, New York.

Chapter 9

Specification and Construction of Performability Models

John F. Meyer
William H. Sanders

Model-based performability evaluation of computer and communication systems requires accurate and efficient techniques for model construction as well as model solution. Moreover, as the physical and logical complexity of such systems continues to grow, there is need for increased care in specifying just what is to be constructed in response to the aims of a given evaluation study. This presentation thus focuses on specification and construction of performability models, under the assumption that construction (and subsequent solution) are automated. Concepts and methods are described for each, with emphasis on specifying/realizing the relation between a base stochastic model and the measures it must support. Although stochastic activity networks are chosen as the vehicle for base model specification, many of the techniques employed convey principles that apply as well to other specification constructs.

9.1   Introduction

THE presentation that follows concerns model-based evaluation of computer and communication system performability, with emphasis on how models for this purpose are specified and constructed. Since contemporary systems of this type are seldom autonomous (closed), such modelling efforts must generally consider the representation of a *total system* composed of

1. an *object system*: the system that is the object of (model-based) evaluation, and

2. an *environment*: other systems (physical or human) that interact with the object system during its use.

Given an interacting set of physical and human resources, the distinction between these two things (although often not made explicitly) depends on which subset of resources is having its ability to perform investigated. This subset comprises the object system, which, in this context, is often referred to simply as the "system". The environment then comprises those remaining resources that interact with the object system to an extent that affects its performability in some appreciable sense.

Accordingly, the above distinction is actually determined by just what aspects of a total system's structure and behaviour are to be assessed via the evaluation process. In this regard, the discussion that follows presumes that such evaluation concerns an object system's "quality" (effectiveness) as opposed to its "cost". Given this qualification, we make some further distinctions that conform with current use of terminology in the computing field. With respect to a designated user-oriented or system-oriented service, *performance* usually refers to some aspect of service quality, assuming the system is correct. In other words, quoting [141], performance is generally indicative of " ... how well a system, assumed to perform correctly, works". (This use is not uniformly adhered to, however; e.g., the text by Kant [254] treats dependability measures as a special class of performance measures.) Performance evaluation and, specifically, model-based evaluation of measures such as throughput, response time and resource utilization, have long been recognized as important in the context of computer and communication system design. Moreover, in the development of theory and techniques for this purpose, there has been remarkable progress over the past 20 years, particularly with regard to extensions and applications of queueing network models.

Although historically an equally long-lived concern, the need to evaluate effects of incorrectness in this context has received less attention. More specifically, we are speaking of incorrect behaviour due to either *design faults* (mistakes made by humans or automated tools in the process of specifying, designing, implementing or modifying a system) or *operational faults* (physical or human-made faults that subsequently occur during system operation). Both types of faults (see [273] for a more thorough treatment of the distinction) can obviously affect a system's ability to perform in a designated environment.

Certain measures of a system's ability to perform are based on the generic concept of *dependability*, i.e., the property of a system that allows "reliance to be justifiably placed on the service it delivers" (again see [273]). Measures of dependability thus quantify an object system's ability to perform with respect to some agreed-upon specification of desired service, where a *failure* of the (object) system occurs when delivered service no longer complies with this specification. Special attributes of dependability are defined according to the nature of failure occurrences and/or their consequences. These include *reliability* (continuity of failure-free service), *availability* (readiness to serve), *safety* (avoidance of catastrophic failures), and *security* (prevention of failures due to unauthorized access and/or handling of information).

However, if performance is degradable, then, as has been well documented in the literature (beginning with [314, 315]), measures of *performability* are needed to address issues of both performance and dependability simultaneously. Specifically, with respect to some designated aspect of system quality, a performability measure quantifies how well the object system performs in the presence of faults over a specified period of time. (This includes special cases of a single time instant at one extreme and an unbounded period of time at the other.) Such measures can thus account for degraded levels of performance that, according to failure criteria, remain satisfactory. They also permit simultaneous (i.e., within the same model) consideration of distinctions among users with respect to how failures and their consequences are perceived.

Given a specification of the measures of interest, a stochastic process or simulation model must represent the object system and environment in sufficient detail to "support" measure solution. A model with this property is thus referred to as a *base model* of the total system. In other words, based on a model's probabilistic nature, a user is able to determine the value or values of each specified measure. With regard to the object system, such modelling considerations are fairly well understood. On the other hand, the importance of realistic environment modelling is often underestimated, particularly when different aspects of the environment have differing effects on various components of the object system. Moreover, if faults are a concern, an environment model needs to account for more than just externally imposed workload, e.g., events such as transient fault occurrences, conditions such as temperature and humidity, and actions of external systems (either physical or human) in effecting fault repair and recovery. In certain cases, it is possible to view the object system as autonomous (relative to the measures in question). For example, much of traditional structure-based reliability evaluation presumes autonomy of this type, meaning that the environment part of the total system model is null. In most contemporary applications, however, the environment is nontrivial. In such cases, its modelling calls for the same kind of care and attention that is typically devoted to the specification and construction of an object system model.

In the evolution of performability evaluation (see [319, 320], for example), the initial emphasis was on solution methods (see [106, 463] for comprehensive surveys of such techniques). However, as a consequence of the above considerations, problems encountered in the specification and construction of performability models have become equally challenging from a technical point of view. Attention to this problem has increased considerably over the past decade (see the recent overviews of [106, Sec. 3] or [217], for example). In addition to the factors cited in the previous paragraph, emphasis on specification/construction issues has also been motivated by the great increase in complexity of the type of (total) systems being evaluated. In particular, with such growth there is a greater need to "match" a model to a few measures of interest, rather than attempt construction of a model that can support a host of measures. In the analytic case, the latter may require a state space that is infeasibly large, thus precluding its actual construction. On the simulation side, even if construction is possible, the details required to support a multitude of measures may result in solution times that, for any given measure, are impractically long.

Prompted by these concerns, the intent of the discussion that follows is to describe, both conceptually and methodologically, what is meant by model "specification" and model "construction" in the context of performability evaluation. The nature of the presentation is tutorial in its style, with the hope of providing a better understanding of both endeavours. This is done via descriptions of generally defined concepts and techniques, followed by some illustrative, concrete examples. The latter presume a particular scheme that employs activity networks (SANs) for specification purposes and the software tool *UltraSAN* for model construction.

In general, the distinction between specification and construction is somewhat arbitrary. Stated informally, we take the view that model *specification* (in the "action" sense of this word) provides the "input" needed to construct and solve a model. Model *construction* is then the act of realizing the specified base model (analytic, simulation, or possibly a combination of both types) that, in turn, supports solution of the specified measures. We assume further that specification, in this context, is essentially a human activity, where the specifier(s) are sufficiently familiar with both (i) the total system being evaluated and (ii) the capabilities of the people and tools responsible for the construction and solution phases. Further, we assume that these phases are realized by a single software tool, thus permitting a more easily defined interface between model specification on the one hand, and model construction/solution on the other. Although this is admittedly a somewhat specialized form of the general problem, its consideration encompasses almost all of the important technical issues that arise in more generally defined settings.

The ensuing discussion is organized accordingly, with Section 9.2 devoted to performability model specification and Section 9.3 to the construction of models so specified. Although performability solution methods lie outside the scope of the discussion, the determination of what is to be solved (by such methods) is an essential part of the specification and must be accounted for during construction. Thus, material in Sections 9.2 and 9.3 relies on certain assumptions regarding solution capability. Section 9.4 concludes the presentation with a summary and some pointers to applications of the methods described herein.

9.2   Performability model specification

A specification of a performability model can be regarded as having three major ingredients.

S1. Specification of what is to be learned about the object system from its (model-based) evaluation, i.e., the performability measures of interest.

S2. Specification of a stochastic process on which the evaluation is to be based (a base model of the total system).

S3. Specification of how S2 relates to S1 in a manner that permits the base model (after construction) to support solution of the specified measures.

Moreover, given that the recipient of the above is a model-based evaluation tool, languages used to state S1–S3 must be sufficiently formal to permit their unambiguous interpretation and subsequent automated realization by the tool.

  A measure, as this term applies to computer and communication system evaluation, typically refers to some aspect of (object) system quality, e.g., throughput, response time, time to failure, etc. However, the manner in which this aspect is measured (in a probability-theoretic sense) is typically implied by the nature of the model, e.g., a performance measure that is based on a queueing model usually refers to the expected value of some random variable under steady-state conditions. However, in the more general setting of performability measures, it is convenient to specify both the aspect of interest and the precise way it is to be measured.

  Using terminology and notation of the modelling framework introduced in [314, 315], S1 thus takes the form of one or more random variables Y together with a specification of how each is to be measured. Such variables are generally referred to as *performance variables* (alternatively *performability variables*), where specification of a particular Y includes

(a) an interval of time (or an instant of time in the degenerate case) over which object system quality is being observed, and

(b) a set A in which Y takes its values (referred to in [314, 315] as the *accomplishment set*).

Specific interpretations of (a) and (b) express the intended meaning of Y and, in turn, what is to be learned about the object system via specified measures thereof. The latter can range from a complete quantification of performability, as supplied by the probability distribution function (PDF) of Y, to single-number measures such as moments of Y or, for a specified set B of accomplishment levels ($B \subseteq A$), a single performability value $Perf(B) = P[Y \in B]$. In what follows, the combination of a specific Y and its specified measure will be referred to simply as a Y-*measure*. When unqualified, the term "measure" will subsequently have the more specific meaning, e.g., it refers to an arbitrary Y-measure or, in contexts where Y is understood, a particular Y-measure.

Regarding S2, we assume that the base model being specified is a discrete-state stochastic process $X = \{X_t \mid t \in T\}$, where the index set (time base) T is continuous (typically the real interval $[0, \infty)$). For any $t \in T$, the value of the random variable X_t represents the state of the total system at time t. Note that X, once constructed, may be characterized in a form that is not literally a stochastic process, e.g., a computer program that simulates X. For specification purposes, however, the more general view of what is being specified is advantageous since, among other things, it allows the same specification to be used for both analytic and simulation model construction.

S3 is perhaps the most difficult aspect of performability model specification, particularly if the techniques employed must apply to different types of evaluation (including strict performance and dependability as well as performability) and, accordingly, a wide variety of specific base models and Y-measures. Its purpose is to specify how state trajectories (sample functions) of X map to values of Y, thus permitting solutions at the base model level (e.g., state occupancy probabilities) to determine the desired value(s) of a Y-measure. (Such a mapping is referred to in [314, 315] as a *capability function*.) Moreover, S3 should rely only on knowledge that is explicit in specifications S1 and S2 as opposed, say, to details that are revealed once the construction of the base model is completed. Although use of the latter might be possible, it is discouraged by two considerations. First, there is no guarantee that the resulting base model can support evaluation of the Y-measure(s) in question. Second, even if support is possible, X may be too complicated when fully constructed (e.g., have too many states) to be dealt with effectively in this regard (i.e., too difficult

for novice or less technical users, and too cumbersome for technical users).
Hence, our development assumes that S3 is specified directly in terms of
S1 and S2.

The overall specification, in addition to reflecting the above considerations, must be such that the subsequent construction phase is indeed
feasible, e.g., the base model X can be characterized within practical limits
of computer memory and compile time. In addition (although solutions
are not an explicit concern of the material that follows), the complexity
of the constructed model should allow feasible and, one hopes, efficient
means of solving the specified Y-measures. Details concerning S1–S3 are
described in the three subsections that follow.

9.2.1   Measure specification

If measure specification is to have general applicability, it must be done
in a manner compatible with a general means of relating, per S3, the base
model specification to the measure specification. To accomplish this, it
is advantageous to view the accomplishment sets A of all performance
variables Y as expressing value with respect to a common, uninterpreted
unit of measure. In a stochastic process setting, a unit of this sort is
typically referred to as a unit of "reward" (see [236], for example). Given
that elements of A express reward, then, quite naturally, Y can be referred
to alternatively as a *reward variable* (see [412], for example).

With this unified view of accomplishment, almost any aspect Y of object system performance (quality) can then be represented by giving reward
a more specific interpretation. Moreover, by our earlier remarks concerning Y specification (see (a) and (b) at the outset of Section 9.2), it remains
only to specify formally the time instant at which, or time interval over
which, reward is being quantified by Y. Here, in concert with the way time
is represented in the base model, time instants and durations associated
with Y can either be elements of the real interval $[0, \infty)$ or, given that
Y has a limiting distribution, be the limit as a time instant or interval
duration approaches infinity. Specifically (again as discussed in [412], but
without presuming an already-specified reward structure), three categories
of reward variables can be usefully distinguished according to the nature
of this time specification.

Reward variables in the first category, called *instant-of-time* variables,

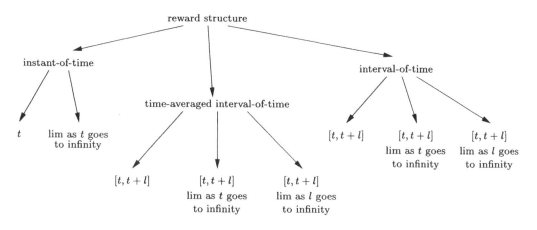

Figure 9.1: Variable specification

represent the reward experienced at a designated time during the object system's use. Variables in the second category, referred to as *interval-of-time* variables, represent the total amount of reward accumulated over a specified interval of time. Variables in the third category, called *time-averaged interval-of-time* variables, are similar to those in the second except that accumulated reward is now averaged over the duration of the specified interval. These three categories are at the first level of the tree depicted in Figure 9.1.

Interval-of-time variables can then be further classified according to the nature of the interval, again as shown in Figure 9.1. The first type represents the total or time-averaged total reward accumulated during some bounded interval $[t, t + \ell]$ (the leaves labelled $[t, t + \ell]$ in Figure 9.1). The second is obtained as a limiting version of the first (the middle leaves of the middle and right branch), where the duration ℓ remains finite and t goes to infinity. This type is useful in representing ability to perform over a bounded period when the system is initially operating under steady-state conditions. The final type (rightmost leaves) is similar to the second except that the initial instant t is fixed and the duration ℓ goes to infinity. For any such variable Y, the specification of a Y-measure is completed by a statement of how Y is to be measured in a probability-theoretic sense. The manner in which reward is interpreted for Y will typically determine which measures will make sense. The following example of a small total system illustrates some possible choices. It is complex enough to exemplify much of what is discussed throughout the remainder of the presentation.

Example: Faulty multiprocessor   Consider an object system consisting of N processing elements (PEs) that share a common input queue of finite capacity L. The environment is a serial stream of incoming tasks that arrive as a Poisson process with rate λ. Whenever queue capacity permits, an arriving task enters the object system; if the queue is full, an arriving task is rejected. Tasks are scheduled to processors on a FIFO basis, so that any "available" processor (the meaning of this will be explained in a moment) will attempt to access the task at the head of the queue. It is assumed that all of the available processors are equally likely to receive the task.

All PEs have identical structures and behave as follows. A PE, having accessed a task when idle, processes that task in exponentially distributed time with mean value $1/\mu$. Moreover, while busy with a single task, it can access a second task and accommodate both as if they were one, i.e., the two are processed together at the single-task rate μ and complete simultaneously. Accordingly, we regard a PE as being *available* if it is either idle or processing a single task; otherwise it is *unavailable* in the sense that additional loading is precluded.

In the case of double-task processing, however, all is not perfect. Specifically, there is interference, due to a fault in the design, that results in processing errors. Fortunately, these are detected at the time both tasks complete, with the likelihood of such errors being encountered given by the probabilities

p_1 = the probability that exactly one of the two tasks
          is processed erroneously,
p_2 = the probability that both tasks are processed erroneously,

where $0 \leq p_1 + p_2 \leq 1$. If a task completes with erroneous processing (we call this an *erroneous completion*), it is immediately returned to the PE for reprocessing. Accordingly, since both tasks may have erroneous completions, we assume further that $p_2 < 1$, so as to eliminate the possibility of live-lock. When the processing of a task completes without errors (an *error-free completion*), which may require several processing iterations, it departs the object system. Thus, any task that enters the system (i.e., is not rejected at the input) will be eventually accessed by some processor and, in turn, will eventually enjoy an error-free completion. In the case in which only a single task resides in a PE during an iteration (either a task that was just accessed or one that was returned internally for reprocessing), everything works fine, and hence error-free completion is guaranteed.

For the total system just described, there are a number of Y-measures that might be considered. For example, the probability that the queue is full at some given time t is of obvious interest, due to the finite capacity of the queue and the slowdown due to reprocessing of tasks with erroneous completions. This measure could be specified, for example, in terms of an instant-of-time variable V_t, where reward is interpreted as the number of tasks in the queue. (Note that, although the reward variable part of the specification is referred to generically as a Y-measure, we take the liberty of using other symbols such as V and W; in particular, the symbol V is intended to suggest an instant-of-time variable.) Hence, when coupled with the interpretation of t,

$$V_t = \text{the number of tasks in the queue at time } t.$$

Recalling that L is the capacity of the input queue, if we take the (singleton) set $B = \{L\}$ to be the accomplishment set of interest, the associated measure is then the performability value

$$Perf(B) = P[V_t = L].$$

Alternatively, this probability could be specified via the binary-valued reward variable (indicator variable)

$$V_t = \begin{cases} 1 & \text{if there are } L \text{ tasks in the queue at time } t \\ 0 & \text{otherwise} \end{cases}$$

coupled with its expected value (mean) as the associated measure. In other words, $E[V_t]$ (by virtue of properties of both V_t and E) likewise expresses the probability of a full queue.

To specify the steady-state probability of the queue being full, one instead uses a variable V_∞ whose PDF is the limiting distribution (provided it exists) of its corresponding V_t variable, e.g., for the first of the two variables considered above,

$$V_\infty = \text{the number of tasks in the queue as } t \to \infty.$$

Then $Perf(B)$, where again $B = \{L\}$, gives the desired measure. Since tasks are assumed to arrive as a Poisson process, $Perf(B)$ also specifies the steady-state probability that an arriving task will encounter a full queue and hence not enter the system.

Measures of productivity of an object system can typically be specified using interval-of-time variables. For example, relative to a particular processor, if a unit of reward is associated with each error-free task completion, then the interval-of-time variable $Y_{[0,\ell]}$ represents the number of error-free completions (for that processor) during the time interval $[0, \ell]$. If the rate of error-free completions, as averaged over $[0, \ell]$, is also of interest, this can be specified via the time-averaged interval-of-time variable

$$W_{[0,\ell]} = \frac{Y_{[0,\ell]}}{\ell},$$

where $Y_{[0,\ell]}$ is as above. Corresponding steady-state variables are obtained by considering their limits as $\ell \to \infty$. Performability measures associated with steady-state variables could range from simple mean values to full probability distributions.

9.2.2   Base model specification

Let us turn now to the second ingredient, namely the specification of a stochastic process X on which the evaluation (solution of the measures given by S1) is to be based. One means of doing this, of course, is to specify X directly, e.g., in the case of a finite-state, time-homogeneous Markov process, via a specification of X's initial-state distribution and its infinitesimal generator. However, the following discussion is aimed at higher-level specifications from which a base model X be can automatically constructed (by a tool that receives the specification).

Stochastic activity networks   The most popular vehicle for accomplishing performability evaluation has been forms of stochastic Petri nets (SPN) [328, 349], which are typically referred to as "models". However, as we use them in the context of S2, such graphical models serve to specify lower-level stochastic models (base models); hence, when there is need to emphasize this distinction, we will refer to these graphical models more precisely as model specifications. Also, due to space limitations, we choose to focus on an SPN-variant that is most familiar to us, namely stochastic activity networks (SANs) [333, 318, 405]. Finally, in keeping with this choice, the tool we use for construction purposes is UltraSAN (see [96, 415]). Among other things, this permits discussion and illustration of hierarchical specification/construction techniques that are not implemented in an earlier SAN-based tool (Metasan [409]).

Structurally, SANs have primitives consisting of *activities, places, input gates* and *output gates*. Activities ("transitions" in Petri net terminology) are of two types, *timed* and *instantaneous*. Timed activities represent actions of the object system or environment whose durations impact on the measures in question. Instantaneous activities, on the other hand, represent actions that, for modelling purposes (support of the measures), can be regarded as having negligible durations. Cases associated with activities permit the specification of two types of spatial uncertainty. Uncertainty about which activities are enabled in a certain state is specified by cases associated with intervening instantaneous activities. Uncertainty about the next state assumed upon completion of a timed activity is specified by cases associated with that activity. Places are as in Petri nets and may contain *tokens*. An assignment of numbers of tokens to places in the network is a *marking* of the network. Input gates each have an *enabling predicate* and *function* that, as will be seen below, control the execution of the network. Output gates have only *functions*, which define the change in marking of a network upon completion of activities. The use of gates permits greater flexibility in specifying enabling and completion rules than with ordinary stochastic Petri nets.

The stochastic nature of a SAN is described by associating an *activity time distribution function* with each timed activity and a *probability distribution* with each set of cases. Generally, both distributions can depend on the global marking of the network. The activity time distribution can be any probability distribution function and is dependent on the marking in which the activity is "activated" (see below). The probability distribution associated with cases, referred to as the *case distribution*, can depend on the marking of the network at completion time of the associated activity.

Before describing how a SAN executes in time, it helps to define a few related terms. In particular, a *stable* marking of a SAN is one in which no instantaneous activities are enabled. Conversely, a marking in which there is at least one instantaneous activity enabled is *unstable*. An activity is *enabled* if the predicate of each of its input gates is true (*holds*, in SAN terminology), and there is at least one token in each of the directly connected input places (i.e., those places connected by a directed arc from the place to the activity).

Informally, SANs execute in time through completions of activities that result in changes in markings. Activities complete some period of time after they are *activated*, depending on their activity time distribution functions. (Activation of an activity occurs when the activity becomes enabled or when it completes while remaining enabled.) More specifically, an activity is chosen to *complete* in the current marking based on the relative priority among activities (instantaneous activities have priority over timed activities) and the activity time distributions of *enabled* activities. Selection of a case of the activity chosen to complete is then based on the probability distribution for that set of cases. These two choices uniquely determine the next (stable or unstable) marking of the network, which is subsequently obtained by executing the input gates connected to the chosen activity and the output gates connected to the chosen case.

Activities may also be restarted, or *reactivated* [318], under certain circumstances. In particular, for each marking in which an activity may be activated, a set of *reactivation markings* can be defined. If, prior to completion, one of these markings is reached, then the activity is *reactivated*, i.e., aborted and then immediately activated. This provides a mechanism for restarting activities, with either the same or a different activity time distribution. One can think of this mechanism as a generalization of the execution policies proposed for other forms of stochastic Petri nets, specified on a per-activity basis.

SAN specification of a single processor   To illustrate the use of stochastic activity networks as specifications, let us again consider the faulty multiprocessor described in the previous section. We are interested in specifying an analytic base model that can support solution of measures of the type illustrated earlier in this section. Figure 9.2 depicts the graphical part of a SAN specification for a one-processor version of the faulty multiprocessor. In the figure, *arrival*, *access* and *processing* are timed activities. *Arrival* and *access* each have a single case, and *processing* has three cases. *Capacity* and *available* are input gates, and *correct* is an output gate. *Size*, *queue* and *num_tasks* are places.

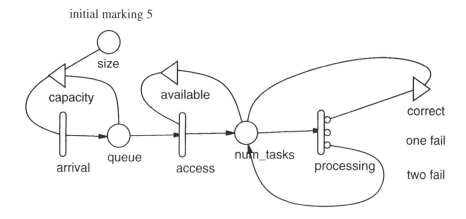

Figure 9.2: SAN specification

Table 9.1: Activity time distributions

Activity	Distribution	Parameter values
access	exponential	
	rate	*1000*
arrival	exponential	
	rate	*1.5*
processing	exponential	
	rate	*1*

This SAN diagram, together with its activity time distributions (Table 9.1), case probabilities (Table 9.2) and input and output gate definitions (Tables 9.3 and 9.4, respectively), precisely specify the object system, workload and fault environment described earlier. To see this, and to illustrate the use of SANs in specifying performability models, we now describe the functioning of the model. In particular, incoming task arrivals are modelled by completions of activity *arrival*. Upon arrival, a task is represented by a token in place *queue*, and the number of tokens in *queue* is the number of arrived tasks waiting for service. The finiteness of the queue is represented by input gate *capacity*, which holds whenever the queue is not full. This can be seen from the gate's predicate in Table 9.3, which specifies that the number of tasks in the queue must be less than the sys-

Table 9.2: Activity case probabilities

Activity	Case	Probability

processing	1	*if ($MARK(num\_tasks) == 1$)*
		       return(1.0);
		   else return(0.81);

	2	*if ($MARK(num\_tasks) == 1$)*
		       return($ZERO$);
		   else return(0.18);

	3	*if ($MARK(num\_tasks) == 1$)*
		       return($ZERO$);
		   else return(0.01);

Table 9.3: Input gate definitions

Gate	Definition
available	<u>Predicate</u> $MARK(num\_tasks) < 2$

	<u>Function</u>
	   / do nothing */*
	   ;

capacity	<u>Predicate</u>
	   / has the buffer capacity been reached? */*
	   $MARK(queue) < MARK(size)$

	<u>Function</u>
	   / do nothing */*
	   ;

Table 9.4: Output gate definitions

Gate	Definition
correct	*/* complete all tasks at the same time */* $MARK(num\_tasks) = 0;$

tem capacity, for the attached activity to be enabled. When an activity completes, a case is chosen (for activity *arrival*, there is only one case, so it is chosen by default), the input gates of the activity are executed, one token is subtracted from each input place, the function of each output gate connected to the chosen case is executed, and one token is added to each output place of the chosen case. Activity *arrival*'s completion thus results in the execution of the function of input gate *capacity* (which does nothing, according to its specification) and the addition of one token to place *queue*, because of the output arc from the activity to place *queue*. The rate of activity *arrival* specifies the rate of task requests, which may or may not be serviced, depending on the number of tokens in place *queue*. Attempts of the PE to access a task from the queue are represented by activity *access*, whose activity time (see Table 9.1) represents the time to acquire a task from the queue, given that the processor is available (i.e., the PE is idle or processing a single task). The enabling of this activity is therefore controlled by input gate *available*. Activity *processing* represents the processing time of the PE. Specifically, when processing completes, the action taken depends on whether one or two tasks were processed. This choice is reflected in the cases of activity *processing* (see Table 9.2), whose probability distribution depends on the marking of place *num_tasks*. If only one task was processed, processing was correct with probability 1, and case 1 (the topmost case, in the diagram) is always chosen. When this occurs, one token is removed from *num_tasks* because of the input arc from the place, and output gate *correct* is executed.

If the processing of two tasks was completed (simultaneously), the outcome is probabilistic, as per the informal specification given earlier. In this situation, the three cases of activity *process* correspond, respectively, to the following alternatives.

1. Both tasks were processed correctly and hence depart the system.

2. One task was processed correctly, thus departing, while the other must be reprocessed.

3. Both tasks were processed incorrectly, and hence both must be reprocessed.

The probabilities associated with each of these actions are given in Table 9.2, and the actions are carried out by the arcs and output gate connected to the activity. For example, if two tasks complete processing, with probability 0.01 they were both processed incorrectly. If this occurs, the third case of the activity is chosen, and the token that was removed from place *num_tasks* by the input arc attached to *processing* is returned via the output arc attached to the case.

The example just presented illustrates the specification of a single processor system as a SAN. While multiple processor versions could be specified directly as SANs through the use of additional activities, places and gates, this would be cumbersome for large systems. Furthermore, structures (such as symmetries) in the SAN that could be exploited in the construction process would be difficult to detect and use profitably. Composition methods for SAN specifications address both of these concerns and are now discussed.

Composed models   As pointed out in the previous paragraph, large systems are cumbersome to express directly in terms of a single SAN specification and can lead to models whose solution is difficult, due to the complexity of the resulting specification. Composed model specifications permit hierarchical composition of SAN models, and their corresponding reward variable specifications (to be discussed in the next section), in an iterative manner. This composition is done using two operations: *replicate* and *join*. The composition acts on the reward structure(s) of a SAN, as well as the SAN itself, to preserve properties needed in the subsequent construction process. Formally, the resulting specification is known as a *composed SAN-based reward model* (composed SBRM) [413].

The *replicate* operation replicates a SAN and associated reward structure a certain number of times, holding some subset of its places, called its "distinguished" or "common" places, common to all resulting submodels. Replicated submodels interact through these common places; although the submodels are identical in their specifications, each of them has its own marking. Each replica will have the same reward structure(s) (see the next section for their specification) as the original submodel.

The *join* operation allows the combination of several different submodels. Informally, the effect of the operation is to produce a composed model that is a combination of the individual submodels. Again, distinguished places play an important role in the operation. In this case, however, a *list* of places is associated with each component submodel. The first places

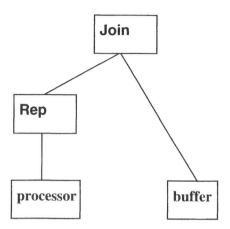

Figure 9.3: Composed N-PE faulty multiprocessor specification

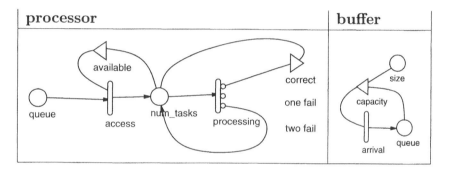

Figure 9.4: SAN submodels in composed model specification

in all of the lists are merged to form a single place, the second places are merged to form another place, and so on. Particular elements on the lists can be null, making it possible to create certain places from a proper subset of the joined submodels.

Composed model for N-PE faulty multiprocessor A composed model specification of an N-PE faulty multiprocessor is given in Figure 9.3. The nodes at the leaves of the tree are SANs, together with their reward structures. These SANs represent the workload offered to the multiprocessor (**buffer**) and the processing at a single PE (**processor**), as shown in Figure 9.4. Note the similarity with the single-processor version described earlier. Since we now consider an N-processor version, we must replicate the processor specification N times. Replication is done via the "**Rep**" node in Figure 9.3. The **Rep** specifies that its child node (**processor**) should be replicated N times, taking *queue* to be a common place for all the replica submodels (again see Figure 9.4). In other words, this operation creates a specification consisting of N processors, all sharing a common place *queue*. Specification of the model is completed by joining the N processor specification to the **buffer** submodel, using the **Join** node in Figure 9.3. This node specifies that the place *queue* in the N-PE submodel should be common with place *queue* in the **buffer** submodel.

9.2.3 Reward structure specification

We now describe how a SAN specification of a base model is linked to measure specifications of the type described in Section 9.2.1. This connection is provided by a *reward structure*, which is similar to that used for Markov reward models, but specified at a level that coincides with the base model's specification (the *SAN level*). A reward structure typically consists of two types of rewards: an *impulse* reward that is associated with each state change and a *rate* reward that is associated with the time spent in a state. This idea was extended in [412] to permit its formulation at the SAN level, where impulse rewards are assigned to activity completions and rate rewards are assigned to particular numbers of tokens in places. A similar extension was made by Ciardo *et al.* [77] in defining "stochastic reward nets".

To describe such structures more formally, let $I\!N$ denote the set of natural numbers and let $\mathcal{P}(P, I\!N)$ be the set of all partial functions from P to $I\!N$. Then, as in [412], an *activity-marking-oriented reward structure* of a SAN, with places P and activities A, is a pair of functions

$$\mathcal{C}: A \to I\!R,$$

where, for all $a \in A$, $\mathcal{C}(a)$ is the reward obtained due to completion of activity a, and

$$\mathcal{R}: \mathcal{P}(P, I\!N) \to I\!R,$$

where, for all $\nu \in \mathcal{P}(P, I\!N)$, $\mathcal{R}(\nu)$ is the rate of reward obtained when, for each $(p, n) \in \nu$, there are n tokens in place p. In this context, an element $\nu \in \mathcal{P}(P, I\!N)$ is referred to as a *partial marking* of the SAN, signalling that this marking refers only to places in a subset of P, namely the domain of the partial function ν.

Informally, impulse rewards are associated with activity completions (via $\mathcal{C}$), and rates of reward are associated with numbers of tokens in sets of places (via $\mathcal{R}$). We adopt the convention, in practice, that rewards associated with activity completions and partial markings are taken to be zero if they are not otherwise explicitly assigned. Performance, dependability and performability variables can then be easily defined in terms of these rewards, using the measure specification method discussed earlier in this section (see Figure 9.1).

Reward structure specification for faulty multiprocessor   To illustrate the use of reward structures, consider those described in Table 9.5 for the composed N-PE faulty multiprocessor. Following the convention employed in *UltraSAN*, we specify rate rewards using multiple predicate-function pairs. The interpretation of each predicate-function pair is as follows. When the predicate is true, reward is earned at the rate specified by the function. The total rate at which reward is accumulated by a variable is then the *sum* of the rewards contributed by all of the predicate-function pairs. Furthermore, from the nature of the replicate operation described in the previous subsection, a reward structure is replicated along with its SAN. The contribution to the total reward by the replicated SAN is thus obtained by summing the rewards associated with all of the replicates.

For example, consider Table 9.5's first reward structure, which specifies a reward structure for an indicator random variable. This variable is used to determine *probability non-blocking*, i.e., the probability that the input queue is not full. In that case, reward is accumulated at rate 1 whenever the queue is not full and at rate 0 when it is full. Using this structure for an instant-of-time variable, we obtain the status of the queue either at a particular time t or in steady state. Since it is an indicator variable, its expected value provides the desired non-blocking probability. The second variable in this table illustrates the use of a reward structure to specify the utilization of the N-PE system, where utilization is defined to be the fraction of PEs that are processing at least one task, i.e., if there are k such PEs ($0 \leq k \leq N$), then the utilization is k/N (having value 1 if the system is fully utilized). Since the SAN (subnet) **processor** is replicated, the corresponding reward structure is likewise replicated (via reward sum-

Table 9.5: Reward rate specification for faulty multiprocessor (impulse rewards are not used in the example, hence omitted from the specification)

Variable	Definition
probability non-blocking	
	Rate rewards
	Subnet = *buffer*
	Predicate: $MARK(queue) < MARK(size)$
	Function: *1*
utilization	
	Rate rewards
	Subnet = *processor*
	Predicate: $MARK(num\_tasks) > 0$
	Function: *1.0 / N*
number of tasks in queue	
	Rate rewards
	Subnet = *buffer*
	Predicate: *1*
	Function: $MARK(queue)$
number of tasks in system	
	Rate rewards
	Subnet = *buffer*
	Predicate: *1*
	Function: $MARK(queue)$
	Subnet = *processor*
	Predicate: *1*
	Function: $MARK(num\_tasks)$
completions of processing	
	Rate rewards
	none

ming; see above). Hence, as indicated in Table 9.5, the reward structure specification for *utilization* is the contribution of a single PE (either 0 or $1/N$ according to whether the PE is idle or busy) to the utilization of the entire system.

The third variable in the table illustrates the use of a reward structure whose predicate is "true" (specified as "1" in *UltraSAN*) for all markings, and whose function changes depending on the marking. This structure is defined in terms of the SAN **buffer** and, through its specification, supports an instant-of-time variable Y_t whose value is the number of tasks in the queue at time t (or in steady state if $t \to \infty$).

The fourth variable has a reward structure involving both the **buffer** and **processor** components of the base-model specification, demonstrating the use of multiple predicate-function pairs that relate to different SANs. Because tokens are counted both in place *queue* and in place *num_tasks* (since the latter are summed, this accounts for all the tasks currently being processed by some PE), the instant-of-time variable obtained is the number of tasks in the N-PE faulty multiprocessor system.

The final variable illustrates the use of impulse rewards. In this variable, one "unit" of reward is accumulated each time activity *processing* completes in any of the **processor** submodels. An interval-of-time variable can be defined with this structure to determine the mean, variance or PDF of the number of processor completions in some fixed interval of time. It is also possible to define more complicated variables that make use of both rate and impulse rewards.

9.3 Performability model construction

The construction of a performability model from a given specification can take many forms, depending on the amount of detail required in the base model and the type of solution method employed. Broadly speaking, model solution can be accomplished by either numerical analysis or simulation. In the case of analysis, the model to be constructed is a stochastic process that can support a feasible solution of the specified Y-measure(s). If solutions are to be obtained by simulation, the end product of the construction phase is a discrete-event simulator, which, when executed, produces the desired estimators of the specified measures. In this case, the simulator itself serves as the base model.

With either form of solution, there are many feasible base models for a given performability model specification, differing in the level of detail they preserve relative to the specification. At one extreme, the models can be very detailed, preserving all the details of the base model's specification and supporting any Y-measure that is specifiable in the sense described in Sections 9.2.1 and 9.2.3. At the other extreme, a base model can be the least refined model that feasibly permits a particular specified Y-measure to be solved (by the designated type of solution method). Models at the detailed end of this spectrum are referred to, quite naturally, as *detailed base models*; following the terminology of [413], those near the other end are *reduced base models*. In the remainder of this section, we first describe algorithms that are necessary and common to the construction of both the detailed and reduced varieties. This is followed by a further discussion of construction techniques for each type, along with examples that illustrate their use.

9.3.1   Marking-transition algorithms

Regardless of the type of model construction method employed, there are several algorithms that are common to both detailed and reduced base model construction methods. Generally speaking, these algorithms convert execution of the model's specification (a hierarchical composition of SANs) in terms of its marking behaviour (as discussed in Section 9.2.2) to a less detailed representation that remains appropriate for the intended use of the model. As with most decisions regarding model construction, what is "appropriate" is determined by what we would like to know about the object system. In this regard, examination of the methods employed for the specification of Y-measures (Section 9.2.1) and their corresponding reward structures (Section 9.2.3) suggests that we would like to preserve information regarding timed-activity completions and times spent in stable markings. On the other hand, there is no need to account for particular sequences of instantaneous activities that might complete, or about sojourns in unstable markings, since neither of these occurrences contributes to the value of a reward variable.

Thus, for the purpose of construction, it is enough to consider the possible sequences of timed-activity completions, along with the sequences of stable markings that result from the completions. More precisely, for each stable marking, each timed activity that may complete in that marking, and each case that may be chosen, it suffices to determine the probability distribution of the next stable marking. After the completion of a timed activity and choice of a case, the next stable marking can be reached in one of two ways. If execution of the gates and arcs connected to the chosen case results immediately in another stable marking, then no additional work is required; the probability associated with this possible next stable marking is simply the probability associated with the case. However, if an unstable marking is reached upon execution of the gates and arcs associated with the chosen case, one must address the following two questions.

1. Does there exist a sequence of subsequent instantaneous activity completions and unstable markings such that a stable marking is never reached?

2. If more than one instantaneous activity is enabled, does the probability distribution across next stable markings depend on which instantaneous activity is chosen to complete first?

The answer to the first question determines whether a next stable marking distribution exists for this marking, timed activity completion and case selection. In SAN terminology, if the answer is *no* for all reachable stable markings, activities that may complete in these markings and cases of these activities, we say that the SAN is *stabilizing* [405]. Unfortunately, this question is undecidable, i.e., there is no decision algorithm for deciding whether a SAN, together with a specified initial marking, is stabilizing (again see [405]). While this is disappointing from a theoretical viewpoint, a computer with finite memory is unable to account for arbitrarily long sequences of intervening instantaneous activity completions and resulting unstable markings. In practice, therefore, one simply prescribes an upper bound on the number of subsequent instantaneous activities that will be considered for some SAN in a given initial marking. If all such sequences have lengths less than or equal to this bound, then the SAN is declared (conservatively) to be stabilizing for that initial marking.

If a SAN is stabilizing in the sense just defined, then the second question becomes important. It asks whether the next stable marking distribution is unique or depends on instantaneous activity choices that are made while the next stable marking distribution is being generated. Recall that, with SANs, choices made upon completion of an activity are determined by its cases. This is in contrast, for example, with the GSPN notion of a "random switch" (see [5], for example) for which a probability distribution is specified on completion of one or more instantaneous activities. Cases are therefore advantageous, since they permit a clear distinction between the alternatives that need to be accounted for by the model, and those that are incidental. In keeping with this distinction, we would like the probability distributions for next stable markings to be independent of the completion choices among concurrently enabled instantaneous activities. If this independence holds for all such distributions, we say that a SAN is *well-specified* [405]. Accordingly, a well-specified SAN has a complete probabilistic specification in the sense that its designated activity time distributions and case distributions suffice to describe its probabilistic behaviour completely. Since solution of the specified Y-measures relies on the knowledge of such behaviour, only well-specified SANs are employed for the purpose of base model specification.

Fortunately, it is decidable whether a SAN is well-specified. Moreover, algorithms [405] exist that are constructive in the following sense. If a SAN in some initial stable marking is well-specified, then the algorithm determines the probability distribution of the next stable marking; if it is not well-specified and, moreover, the SAN violates the well-specified property, then the algorithm returns a *yes* answer to Question 2. Informally, determining whether such choices are influential involves enumerating all possible "paths" (i.e., sequences of instantaneous activities and unstable markings) that are traversed before a stable marking is reached. A relation is then defined on this path set, where two paths are related if the choices made among concurrently enabled instantaneous activities are common. It can be shown that this is a compatibility relation on the path set; that is, it is both reflexive and symmetric. However, it need not be transitive, and hence it is generally not an equivalence relation.

Once the maximal compatibility classes have been determined for the relation, the set of possible next stable markings is computed for each class; this is done by listing the stable markings that result from paths in a given class. The probability distribution for that class is then computed by summing, for each resulting stable marking, the probabilities of the paths that terminate in that marking. Finally, the probability distributions so determined for each class are compared. If they are identical, then there is no dependence on the choice distinctions; moreover, the common distribution becomes the next stable marking distribution for the timed activity, case and initial marking considered. If they are not identical, then the SAN is not well-specified.

Given that a SAN is stabilizing, the above algorithm provides a method for moving from stable marking to stable marking, in which at each step, the timed-activity completion that caused the transition is recorded. If the SAN is not well-specified, the algorithm will detect this fact and halt the computation of the state-level representation. Note that this computation can be done on the fly, with no need to record the unstable markings.

Finally, note that these algorithms have recently been generalized to the class of "path-based" reward structures [380]. In these reward structures, impulses can be associated with individual activities or sequences of instantaneous activities, and the well-specified check algorithms must be modified to check that the distribution of impulse reward obtained when transitioning from one stable marking to another is invariant over different activity choices. A similar algorithm, in the context of more standard stochastic Petri nets, has been proposed by Ciardo and Zijal [85].

The next step in the construction is to determine an appropriate notion of state for the target model. This choice determines whether the resulting base model will be detailed or reduced.

9.3.2 Detailed base model construction

Detailed base models are constructed directly from the base model specification, without regard to the performability measures in question. Construction of base models from stochastic extensions to Petri nets is typically done in this manner, where most often (see [217, 5], for example), the state of the base model is taken to be the stable (also known as "tangible") markings of the network. More detailed notions of state, which keep track of the most recently completed timed activity and the stable marking that is reached when that activity completes, can also be considered [413]. Both of these notions of state are "detailed" in the sense that the exact marking of the specifying SAN is preserved in the definition of state for the target base model. In the case of analytic solution methods, this marking becomes part of the state of the resulting stochastic process; if simulation is used, it is the notion of state that is employed for the generated trajectory when the simulator executes. Distinctions between the needs of analysis and simulation are discussed in the paragraphs that follow.

Construction for analytic solution   If a SAN is well-specified and the possible stable markings are taken to be states, then their evolution with time is a stochastic process. More precisely, under these conditions, the *marking behaviour* of a SAN is the stochastic process

$$(R, T, L) = \{(R_n, T_n, L) \mid n \in I\!N\},$$

where T_n is the time of the nth timed-activity completion, R_n is the stable marking reached after the nth timed-activity completion, and L is the total number of state transitions of the process through time T_n (including the one made at T_0), given that $R_0 = \mu_0$ and $T_0 = 0$. Note that since separate random variables are used for the marking entered and the time of entry, the number of activity completions during an interval can be counted. Similarly, activity completions that do not change the marking of the network can be detected, since successive times of activity completions are recorded by T_n. Note that we have made no assumptions about the nature of this stochastic process, which may be Markov, semi-Markov, Markov regenerative, or even more general, depending on the activity time distributions chosen and the structure of the SAN and composed model.

When this amount of detail is not needed, the "minimal marking behaviour" can serve as a detailed base model. More formally, the *minimal marking behaviour* of a SAN with marking behaviour (R, T, L) is the stochastic process

$$Z = \{Z_t \mid t \in I\!R^+\},$$

where Z_t is the stable marking at time t. By ignoring activity completions that leave the marking unchanged, this behaviour is minimal in the sense that it has a minimum number of state transitions relative to the original marking behaviour. Of course, with such simplification, these ignored completions can no longer be detected. For many applications, such detection is not essential; indeed, the minimal marking behaviour is the stochastic process that is typically associated with stochastic Petri nets (see [328, 349, 5], for example).

The resulting marking and minimal marking behaviours are Markov (continuous-time, finite-state, time-homogeneous) if all the activity times are exponentially distributed and, further, activities are reactivated often enough to ensure that their rates (if marking-dependent) depend only on the current state. It is important to note, however, that if the latter condition does not hold, then even if all the activity times are exponential, the minimal behaviour may not be Markov. This possibly surprising fact follows from the execution rules for SANs. In particular, recall that activity times are determined at activation time and may be marking-dependent. Therefore, depending on the nature of the specification, it may be that an exponentially distributed activity time depends on a *past marking* whose rate differs from that of the current marking; if this occurs, then Z is not Markov. Other stochastic Petri net definitions preclude this behaviour, implicitly assuming that rates are determined by the current marking of the net. While this is a reasonable restriction for the purpose of obtaining Markov behaviour (and is specifiable with SANs via reactivation functions), there are other conditions that can ensure that Z is Markov. Furthermore, if all activities are not exponential, it unreasonable to assume that rates "adjust" as markings change, since the delay behaviour of these activities is not memoryless. For additional discussion of this issue, along with a precise definition of the class of SANs that exhibit Markov behaviour, see [405].

Given that the behaviour of a SAN is Markov, a state-transition-rate diagram for its marking behaviour can be determined as follows (see [413] for a more precise description of this algorithm). First, each activity that may complete in the initial state is completed, generating potential new states that correspond to each possible next stable marking that may be reached from the initial marking. If a potential next state already exists, a non-zero rate from the original state to the reached state is added to the list of rates associated with the originating state. If the reached state is new, then it is added to the list of states that need to be expanded. A rate from the original state to the new state is then added to the list of rates for the original state. Generation of the state-transition-rate structure then proceeds by selection of states from the list of unexpanded states and repetition of the above operations. The procedure terminates when (i) there are no more unexpanded states (signifying that the state space is finite and has been completely determined) or (ii) the machine has reached its capacity and cannot store additional states. In the latter case, the state space is either infinite or too large to be computed.

Table 9.6: State-space sizes for detailed base model construction methods

| | Marking behaviour | | | | Activity-marking behaviour | | | |
| | L | | | | L | | | |
N	1	5	10	20	1	5	10	20
1	6	18	33	63	12	44	84	164
2	18	54	99	189	58	214	409	799
5	486	1,458	2,673	5,103	3,483	12,555	23,895	46,575
7	4,374	13,122	24,056	45,972	43,012	292,330	-	-
10	118,098	-	-	-	-	-	-	-

Marking behaviour of the faulty multiprocessor example   To illustrate the marking behaviour associated with a SAN-specified base model, consider again the N-PE example addressed in Section 9.2, with particular reference to the composed SAN specification of its base model (described in Section 9.2.2). To restrict the example to a manageable number of states, let us suppose further that $N = 2$ (the number of PEs) and $L = 1$ (the capacity of the input queue). The result of applying the construction method described above to those choices is an 18-state process having the state-transition-rate diagram depicted in Figure 9.5. In the figure, circles represent possible marking states, and the numbers within the circles refer to the markings of places in those states. The two numbers on the first line refer to the marking of the first processor, the two on the second line the marking of the second processor, and the one on the third the marking of the buffer subnet. The marking of place *size* in the buffer subnet is not shown, since it is constant and equal to the capacity $L = 2$ for all states. The marking for each SAN is such that the places are in alphabetical order. To reduce the complexity of the figure, rates are not shown on the arcs, but they are generated via the algorithm just described. As can be seen from the figure, the marking behaviour for even a small instance of the example system is quite complex. Furthermore, as shown in Table 9.6, the state-space size grows rapidly as the values of N and L increase. Dashes in the table represent state spaces that were too large to generate.

In spite of this fairly rapid growth in size, the marking behaviour is not always sufficient to support a variable that depends on knowledge of *which* activity completed in a particular marking. This is due to the fact

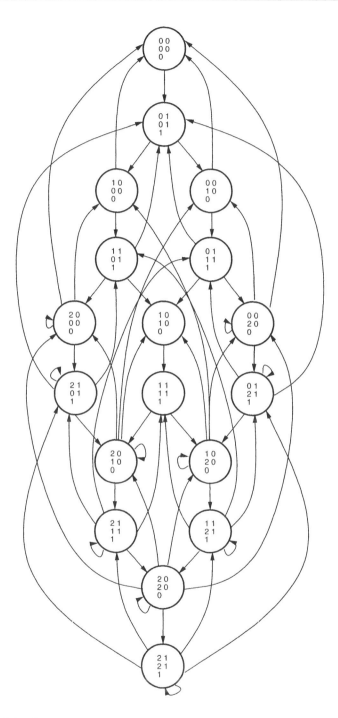

Figure 9.5: Marking behaviour for a 2-PE system with buffer capacity 1

that multiple timed activities may complete in a marking, all resulting in the same next stable marking. These multiple activities induce the same state transition in the marking process and hence cannot be distinguished.

If the marking behaviour is not sufficient, one must keep track of the most recently completed timed activity, as well as stable marking, as part of the state description. The resulting process, called the *activity-marking behaviour* of a SAN [405], is the stochastic process $(R, T, L) = \{R_n, T_n, L\}$, where T_n is the time of the nth timed-activity completion, R_n is the state reached after the nth timed-activity completion, and L is the total number of transitions of the process (including the one made at T_0), given that $R_0 = (\triangledown, \mu_0)$ and $T_0 = 0$. States of the activity-marking behaviour are called *am-states* and are denoted by a pair (a, μ) where μ is a marking and a is a timed activity whose completion can result in μ. The activity-marking behaviour thus distinguishes states with respect to the timed activity that caused a marking to be reached, as well as the marking itself. An am-state is a possible stable marking of the network together with a timed activity that may result in that marking when it completes. $\triangledown$ is a fictitious activity that is assumed to complete, bringing the network into its initial marking.

It can be shown [413] that the activity-marking behaviour of a network is Markov whenever the marking behaviour is Markov, and that the activity-marking behaviour supports all performance variables that can be defined using the reward variable specification method described in Section 9.2. However, this construction method results in base models that become extremely large as the size of a system grows. For example, Table 9.6 shows that the size of the resulting activity-marking space of the faulty multi-processor model grows very rapidly as the number of processors and buffer stages is increased. Even if one considers only the marking behaviour, the state space grows quickly and becomes unmanageable for most realistic applications. This motivates the development of reduced base model construction methods, which we will discuss in the next section. Before doing so, however, we outline a method for simulating SAN-based reward models, using the notions of state just discussed.

Construction for solution by simulation   In a simple approach to discrete-event simulation, each activity is an event type and every activity completion is an event (e.g., [405, 72, 461]). Such an approach must be employed if one wishes to preserve the complete marking or am-state during execution. Simulation of stochastic extensions to Petri nets is typically done in this manner, although simulation of a generated stochastic process representation has also been proposed [137]. Use has also been made of the structural relationship between activities (transitions in GSPNs) [72] to minimize the number of event types that need to be checked for a change in status while retaining the detailed notion of state.

In such a procedure (for example, as implemented in [409]), detailed state trajectories are generated by repeatedly completing the earliest activity scheduled to complete and updating a single future event list, which keeps track of activities scheduled to complete. At every activity completion, a new stable marking for the model is chosen probabilistically from the set of next stable markings generated, using a procedure similar to that discussed in subsection 9.3.1 of this section. It is then necessary to check all scheduled events to see if they are still enabled in the new marking. The events that remain enabled must be kept on the future events list, unless the current marking is a reactivation marking for the activity. If it is, the activity is first removed from the list, and then added back to the list, with the new scheduled completion time. Finally, all activities that are not currently on the future event list must be checked to see if they are now enabled and hence should be added to the list.

While this procedure is simple, it does have some drawbacks. These stem primarily from the fact that as the number of activities (or transitions, in stochastic Petri net terminology) grows, the work that must be done to update the future event list upon each state change also increases. This work relates primarily to the fact that after a change in marking, a large number of activities need to be checked to see if their status (enabled or disabled) has changed. This problem can be solved by using ideas from reduced base model construction and using structural information to minimize the number of activities that must be checked upon each state change [72]. This will be discussed in the next subsection.

9.3.3 Reduced base model construction

Detailed base model construction methods can lead to base models that are extremely large if analytic solution methods are employed, or if simulation is used and a large number of events need to be considered during each state change. To avoid these problems, we make use of the SAN and composed model structure to develop a less-refined notion of state that does not distinguish between replicate submodels. This notion of state is adaptive, and depends on the structure of the composed model tree and choice of performance variables. By exploiting symmetries present in the composed model (identified via the replicate node), the procedures generate base models for analytic solution that often consist of many fewer states than would otherwise be necessary. If simulation is used, the procedures reduce the number of events that must be processed upon each state change. This subsection will first describe the notion of state we employ, which is represented graphically as a "state tree". It will then describe how state trees are used in analytical and simulation-based construction methods.

State trees   State trees [96, 406] are closely related to the graphical representation of composed SAN-based reward models (SBRMs), described in Section 9.2. Recall that a composed SBRM is a result of operations on SBRMs that may themselves be composed models. This composition is represented graphically by a tree. A notion of "state" can then be determined at each level of the tree structure. At a join operation, we keep a vector of "states" for each joined submodel. At a replicate operation, the number of replicas in each existing submodel "state" is kept. Finally, at the lowest level, the "state" of a SAN model is its normal marking. The complete state is then the impulse reward due to the last activity completion and the composed state formed as above. The notion of state thus preserves the identity of all submodels at a join node, but only keeps track of the *number* of submodels in particular states at replicate nodes.

State trees are a graphical representation of the marking portion of a reduced base model state. They consist of three types of nodes: *join nodes*, *replicate nodes* and *SAN nodes*. All leaves of a state tree are of type *SAN*. Nodes that are not leaves are of type *join* or *replicate*. A node of type *SAN* has a *sub-type* that relates the node to a particular SAN model. Each node in the state tree has a corresponding node in the composed model diagram. A node on a particular level on a state tree corresponds in type and level with a node on the composed model diagram. In both the state tree and the composed model diagram, nodes related to SANs are at the leaves of the tree.

Furthermore, each state tree node has associated with it a subset of the distinguished places of the corresponding node in the composed model

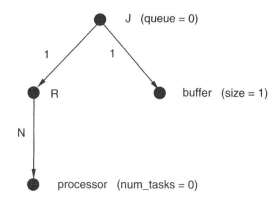

Figure 9.6: Initial reduced base model state

diagram. This subset consists of those places that are distinguished at the node, but not at its parent node. For convenience, we use the expression "a place at node i" to denote this relationship between nodes in the state trees and places in the composed SAN-based reward model. Given this assignment of places to nodes in the state tree, we define μ_i as the restriction of the global marking to the places at node i. The marking μ_i appears next to a node i and is ordered according to the alphabetical order of the places at that node.

Nodes in a state tree are connected by directed arcs. An arc that connects a parent node i to a node j has an associated integer $n_{i,j}$, where $n_{i,j}$ is the number of occurrences of the marking of the SBRM represented by node j at node i. By definition, each outgoing arc j from a join node i has $n_{i,j}$ equal to one, since one copy of each constituent SBRM is used in the join operation. Outgoing arcs from replicate nodes can have cardinality ranging from 1 to the degree of replication defined in the corresponding composed model node, and represent the number of replicas in a particular sub-state.

State trees for the faulty multiprocessor model To illustrate the use of state trees, consider again the N-processor faulty multiprocessor model introduced in Section 9.2. The state tree for the initial reduced base model state of this model, when the number of buffer stages is 1, is given in Figure 9.6. Note that the common place *queue* is at the top level join node (denoted by "**J**" in the figure), since it is common to all processor

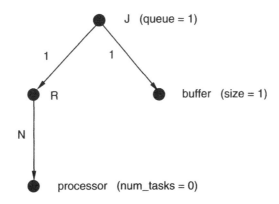

Figure 9.7: Second reduced base model state

replicas and the buffer. No places are at the replicate node (denoted with an "**R**"), since the only place common at that level (*queue*) is also common at the next higher level. The "**N**" on the outgoing arc from the replicate node denotes that all N replicas of the processor are in a single marking and the number of tasks at the processor is zero.

Only one activity, *arrival* in submodel buffer, is enabled in the initial marking. It will therefore eventually complete, resulting in the state tree shown in Figure 9.7. Note that the structure of the tree has not changed, only the marking of place queue at the join node. In this marking, the queue is full, so activity *arrival* is no longer enabled. Activity *access* is now enabled in all processor submodels, which "compete" for the task. Eventually, one of these activities will complete, resulting in the state tree shown in Figure 9.8.

Note that the structure of the tree has now changed, with the processor submodels split into two groups: $N-1$ models that remain in the idle state, and one model that now has a token in place *num_tasks*, indicating that it is processing a single task. It is important to note that the particular processor submodel in which activity *access* completed is not recorded, only the fact that activity *access* completed in some processor submodel. This observation gives insight into the cause of the reduction in number of states in a reduced base model. For a replicate node with N competitively enabled timed activities, this abstraction results in an N-fold reduction of possible next stable markings. If the composed model consists of multiple

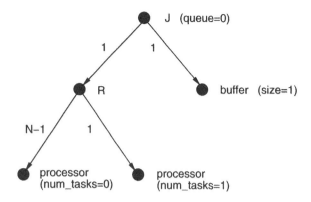

Figure 9.8: Third reduced base model state

replica nodes, arranged in a hierarchical fashion, the reduction can be even greater.

Construction for analytical solution   A state tree is combined with the impulse reward of the most recently completed timed activity to form a *complete state* in a reduced base model. Note that the leaves of the state tree are submodels in a particular marking, and that the number of submodels in that marking can be determined by multiplying together the weights associated with each of the arcs on the path from the root of the tree to the submodel.

A reduced base model then can be generated, conceptually, by taking the reachable complete states as the states in a stochastic process and computing transitions between these states. The process to generate a reduced base model is very similar to that used to generate a detailed base model, except that the algorithm operates on complete states. The rates from one complete state to another are determined as described earlier in this section, except that we consider completions of activities from sets of *replica activities*, which are sets of identical activities in different replicas of a submodel in a particular marking. The SAN will transition from a state when the first activity in some set of replica activities completes, so the departure rate assigned to a set of replica activities in a leaf in the state tree is equal to the rate assigned to that activity in the SAN, multiplied by the number of submodels that are in the marking. Using these rates, and the generated state transition probabilities, transitions occur directly from one reduced base model state to another.

Moreover, it can be shown (see [413]) that the resulting reduced base model has the following properties.

1. It is Markov if the corresponding detailed base model is a Markov process.

2. It supports the specified performance variable.

The first fact is established by formally specifying a mapping from each am-state of the model to its corresponding reduced base model state and showing that this mapping defines a strong lumping on the am-behaviour (activity-marking behaviour) of the model. Reduced base models thus produce exact results and can be solved using Markov methods whenever detailed base models can. It is important to note that while the proof of the Markov nature of the reduced base model relies on lumping arguments, the generation process described above does not. Instead, it generates the reduced base model directly from the composed SAN specification, without ever generating the activity-marking behaviour of the model.

The second fact follows from the variable-specification method described in Section 9.2. Since the impulse and rate rewards are specified in terms of a SAN, they are identical for all replica SANs defined by the replicate operation. Hence, all such SANs have identical rewards when in the same marking. This implies that the reward obtained while executing a composed SAN model depends only on the number of replica SANs in a particular marking, not the fact that a particular replica is in some marking.

Reduced base model for faulty multiprocessor model   To illustrate analytic reduced base model construction, consider once again the faulty multiprocessor introduced in Section 9.2. Figure 9.5 gave the marking behaviour of this model when the number of buffer stages is one and there are two PEs; Figure 9.9 shows the reduced base model for the same model parameters, when all activities have identical impulse rewards. Replica submodels in identical markings are not distinguished (as per the state tree concept), and hence, a state can be labelled by the distinct markings of each SAN, and the number of SANs in each of these markings. As before, markings of SANs are shown as vectors of places in alphabetical order,

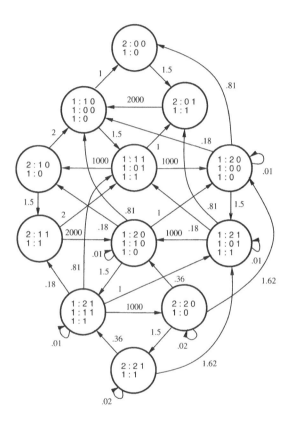

Figure 9.9: Reduced base model for 2-PE system with buffer capacity 1

Table 9.7: Detailed vs. reduced base model construction methods

| | Marking behaviour | | | | Reduced base model | | | |
| | L | | | | L | | | |
N	1	5	10	20	1	5	10	20
1	6	18	33	63	6	18	33	63
2	18	54	99	189	12	36	66	126
5	486	1,458	2,673	5,103	42	126	231	441
7	4,374	13,122	24,056	45,972	72	216	396	756
10	118,098	-	-	-	132	396	726	1,386
15	-	-	-	-	272	816	1,496	2,856
20	-	-	-	-	462	1,386	2,541	4,851
50	-	-	-	-	2,652	7,956	14,586	27,846
100	-	-	-	-	10,302	30,906	56,661	108,171
250	-	-	-	-	63,252	189,756	347,886	-
500	-	-	-	-	251,502	-	-	-

and the marking of place *size* in submodel buffer is not shown, since it is constant for all states. The number of submodels in a particular marking is given by the number before the colon on a line, and the vector after the colon represents the marking of that number of submodels. The numbers on the arcs are rates between states, calculated as described earlier.

The top state in the diagram thus represents the case in which both processors are idle, and the buffer is empty. The state below and to the left represents the situation in which one of the processor submodels has a single token in place *num_tasks*, one processor submodel has no tokens in place *num_tasks*, and there are zero tokens in place *buffer*. Since we require that reduced base models support variables that depend on activity completions, the impulse reward of the activity that most recently completed would normally also be part of the state label; however, since this impulse is the same for all activities, the label is not needed. Note that the reduced base model consists of 12 states, while the marking behaviour consisted of 18 states. The saving comes from the fact that we do not distinguish between the two processor submodels in the reduced base model.

The differences in state-space size are dramatic for larger systems, as shown in Table 9.7. As with Table 9.6, dashes represent state spaces that were too large to generate. As can be seen from the table, generating detailed base models becomes impractical after only 10 processors, but models for up to 500 processors can be generated when reduced base model construction methods are used.

Construction for solution by simulation Recall that when detailed base model construction methods are used, activities in SANs are event types, and activity completions are events in the discrete event simulation. For large models, this leads to situations in which there are a very large number (possibly thousands) of event types, and the number of activities whose "status" (i.e., enabled or disabled) must be checked upon each activity completion is correspondingly large. In simulation, reduced base model construction methods make use of the state tree to perform future event list management more efficiently [406].

This efficiency is achieved by reducing the number of activities that are checked for changes in their status during the transition from one reduced base model state to another. The reduction can be achieved since all replicas of each activity will have the same status, since they all have their input places in the same marking. By definition, this will be the case for all replica activities at a particular leaf in the state tree, and hence we only need to check the status of one activity in a set of replica activities at a leaf in the state tree. More formally, we define a *representative activity* as an activity that "represents" the set of replica activities $a_1 \in A_1$, $a_2 \in A_2$, ..., $a_i \in A_i$, ..., $a_n \in A_n$, where A_i is the set of activities of the ith replica in a set of n replicas of a particular submodel in identical markings. Each representative activity is an *event type* in the new simulation technique, whereas activity completions are events.

During simulation, we operate on representative activities, instead of all activities, when performing future event list management. Status checks on activities are then reduced to a single check per set of replica activities for a set of submodels in identical markings. The events for each of these replica activities can be grouped into a list related to the representative activity. We call this list of sampled completion times a "compound event". More formally, we define a *compound event* e_a for representative activity a as the list of sampled completion times $\{t_1, t_2, \ldots, t_n\}$, where n is the number of activities represented by a. As argued in the previous section, n can be found by multiplying together the numbers on the arcs on the path

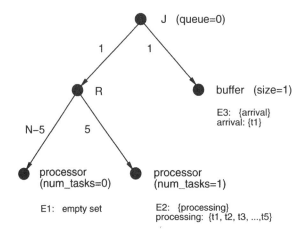

Figure 9.10: State tree with future event lists

from the root to the SAN under consideration. Using this information, it is possible to build compound events, each with n elements, from the list of future events for each set of submodels represented by a leaf node.

Simulation of faulty multiprocessor model   The faulty multiprocessor model is useful for illustrating the use of compound events in simulation using state trees and multiple future event lists. Specifically, consider the state tree of Figure 9.10, which has each leaf node augmented with a list of compound events. This state represents the situation in which five PEs are processing a single task. In this state, there are three sets of compound events: $E1$, $E2$ and $E3$. $E1$ has no events scheduled, since there are no activities enabled in this state in this submodel. $E2$ has one compound event scheduled, which corresponds to activity *processing*. Because the integers at each arc on the route from the node at the highest level to the leaf have been multiplied, there are five replicas of type processor in the same marking; as a result, *processing* has five sampled completion times. $E3$ has one event scheduled, since the queue is now empty; hence, activity *arrival* is enabled. There is only one time scheduled for this event, since the submodel corresponding to this leaf is not replicated.

The algorithms to effect the transition from one state tree (augmented with future events lists) to another are quite complicated, since trees may have multiple replicate nodes, each of which may split or join in a given state change. A detailed description of the algorithms, which also make use of structural information to detect activities that may have changed their statuses, can be found in [406].

9.4 Summary

As argued in the introduction, model specification and construction are important aspects of performability evaluation. With appropriate abstractions and tools, they facilitate the specification of complex behaviours that would be extremely difficult to represent at the state level. Early specification of performance measures is also very important, since they can guide the specification of the environment and object system models. Furthermore, a specification formalism of the type described can have a considerable influence on model construction. In particular, it suggests how base models can be tailored to the measures in question and, by identifying symmetries prior to construction, permits automated reduction of the resulting model. Accordingly, when compared with more conventional approaches, the base models obtained are typically smaller and easier to solve.

While a relatively simple system was used to illustrate some of these advantages, they have all been substantiated through application to a wide variety of more realistic examples. These have served to validate the utility of the specification/construction techniques and, when limits were reached, to motivate additional work. Early (*circa* 1986) techniques of this type were employed by Metasan; subsequently, all the methods described herein have been implemented in UltraSAN. Both tools have been used to evaluate various systems with respect to measures of performance and dependability as well as performability. Examples of the latter include the performability evaluation of computers (e.g., [16, 322]), communication systems (e.g., [15, 321, 375, 294, 255, 379]), databases (e.g., [408, 407]) and software systems (e.g., [295, 453]). The results reveal that stochastic activity networks are indeed an appropriate method for specifying complex system models. Moreover, they have shown that for many practical examples, reduced base model construction techniques result in base models that can be solved using readily available computing resources.

Acknowledgement

This work was supported in part by the Digital Equipment Corporation Faculty Program: Incentives for Excellence and NASA Grant NAG 1-1782.

Chapter 10

A Survey of Performability Modelling Tools

Boudewijn R. Haverkort
Kishor S. Trivedi

Over the last two decades considerable effort has been made in the development of techniques to assess the performability of computer and communication systems. Although these techniques are fundamental to actually evaluating system performability, a prerequisite of a practical but no less important nature is the availability of software tools that support these modelling and evaluation techniques. In this chapter, we specifically address the issue of performability modelling and evaluation tools.

10.1 Introduction

OVER the last 20 years, considerable effort has been made in the development of techniques to assess the performance and dependability of computer and communication systems in an integrated way. This combined performance and dependability modelling, also known as *performability* modelling, becomes necessary when the system under study can operate in a gracefully degradable manner in the presence of component failures. Fault-tolerant computer systems, distributed systems, computer networks and many embedded systems behave in this way.

Although these evaluation techniques are of fundamental importance in enabling a designer to properly engineer a system to satisfy its performance, dependability and performability requirements, a prerequisite of a more practical but no less important nature is the availability of software tools to support performability modelling and evaluation techniques, to allow system designers to incorporate and truly use these techniques when designing systems.

Performability modelling and evaluation tools should be structured such that the models can be specified at a level that is easy to understand for a system designer, and that hides the mathematical details as much as possible. The output of the tool should also be such that it can be understood without detailed knowledge of the underlying mathematical model (where a complete ignorance of the latter is certainly not what we prefer to see).

A number of tools that are in some way suitable for performability modelling and evaluation have been reported in the literature recently. The aim of this chapter is to survey and compare these tools. Aspects that we address are, among others, the modelling language employed, the class of models that can be handled (from a mathematical point of view), and aspects of their user-interface and implementation.

Some of the tools presented in this chapter have been treated by earlier surveys, e.g., by Haverkort and Niemegeers [213], Haverkort and Trivedi [217], Johnson Jr. and Malek [243], Mulazzani and Trivedi [337], Meyer [320, 319], Geist and Trivedi [157], and de Souza e Silva and Gail [106]. However, in this chapter we specifically emphasize performability modelling tools, which was not the case in the papers cited (apart from [213]): Haverkort and Trivedi [217] assess all kinds of formalisms and tools for Markov reward models, i.e., there is no special emphasis on performability; Johnson Jr. and Malek [243], Mulazzani and Trivedi [337] and Geist and Trivedi [157] emphasize tools for dependability evaluation; Meyer [319, 320] describes the historical development of performability, thereby briefly addressing six tools, in far less detail than we do here; de Souza e Silva and Gail [106] emphasize randomization-based performability evaluation techniques, while addressing a few tools, but, apart from the tool TANGRAM, again in far less detail than we do here. This chapter, which can be seen as an extended and thoroughly updated version of [213], provides a systematic overview of 17 performability evaluation tools.

This chapter is organized as follows. In Section 10.2 we discuss a fairly general modelling tool framework that turns out to be useful for the classification of performability evaluation tools. We then briefly touch upon a classification of performability measures of interest in Section 10.3. Then, in Section 10.4, the main part of this chapter, we discuss 17 tools for performability modelling and evaluation. In Section 10.5 we summarize the chapter, point at recent developments and suggest some future areas.

10.2 A general modelling tool framework

As seen from the preceding chapters in this book, there are many evaluation techniques for obtaining specific measures from Markov reward models. Most of these techniques employ numerical algorithms and can therefore only be used when computer support is available. However, even when computer support is available for the numerical solution there are a number of problems that hinder the use of the state-of-the-art Markov reward model solution techniques:

- system designers are in general not familiar with Markov reward models, so they are not inclined to use them to directly model their system;

- Markov reward models are very "low-level" from a system designer's point of view, and thus the modelling process is very error-prone;

- models tend to become too big to handle manually, i.e., the model specification becomes too complex.

In a similar context, Berson, de Souza e Silva and Muntz [34] distinguished two representations of a (performability) model. On one side, there is the *"analytical representation"* of a model. This is the representation that is directly suitable for numerical solution. In the context of performability modelling and evaluation, an analytical representation of a model would just be a Markov reward model and the numerical solution would use one of the techniques discussed earlier in the book. On the other side, there is the *"modeller's representation"* of a model. This is a description in a symbolic form oriented towards the specific application, i.e., the system to be modelled. Clearly, most system designers prefer to use the modeller's representation rather than the analytical representation.

In the performance modelling context the same distinction applies. Beilner claims that a performance model specification used in a performance modelling tool should be independent of any particular performance evaluation technique [29]. This is in line with the statement above that for system designers the analytical representation of a model, which is typically tailored to some specific evaluation technique, is hard to use. Model descriptions should therefore be based on a formalism close to the application domain and as independent as possible from underlying evaluation techniques.

We now define a general framework for performability modelling tools, called GMTF (General Modelling Tool Framework) [201]. Note that although we discuss the GMTF here in a performability modelling context, it has a wider applicability in the context of quantitative systems modelling. We introduce a hierarchy of modelling formalisms ranging from $\mathcal{F}_0$ (the lowest level) to $\mathcal{F}_n$ (the highest level). $\mathcal{F}_0$ yields models that are directly suitable for numerical solution whereas $\mathcal{F}_n$ is the formalism closest to the application domain.

We define $\mathcal{F}_i$-*modelling* as the process of abstracting, simplifying and/or rewriting a system description S in such a way that it fits some formalism $\mathcal{F}_i$. The result of this process is called an $\mathcal{F}_i$-model $\mathcal{M}_i$ of S. An $\mathcal{F}_i$-model $\mathcal{M}_i$ of S can be transformed into another formalism $\mathcal{F}_{i-1}$ ($i \in I\!N^+$), yielding an $\mathcal{F}_{i-1}$-model $\mathcal{M}_{i-1}$ of S. This is called $\mathcal{F}_{i-1}$-modelling. The lowest level formalism is $\mathcal{F}_0$ which coincides with the analytical representation in [34]. When $\mathcal{F}_i$ is the highest level formalism, most of the user activity in the modelling process will be $\mathcal{F}_i$ modelling. The lower level modelling activities will often be partially or completely automated.

Once we have constructed a model, it can be solved. The solution of an $\mathcal{F}_0$-model $\mathcal{M}_0$ yields results in the formalism $\mathcal{R}_0$. For this solution process, most of the techniques addressed in this book can be applied directly, since the formalism $\mathcal{F}_0$ directly suits them. We call the solution of an $\mathcal{F}_0$-model $\mathcal{M}_0$ a $\mathcal{V}_0$-*solution*. The results in the formalism or domain $\mathcal{R}_i$ ($i \in I\!N$) can be further processed or enhanced to some higher level $\mathcal{R}_{i+1}$. This is called $\mathcal{E}_{i+1}$-*enhancement*. Often these enhancements can be done automatically.

When we have an $\mathcal{F}_j$-model $\mathcal{M}_j$ of some system S we want to solve this model and to obtain measures in a formalism of the same level. We denote the level of this formalism as $\mathcal{R}_j$. We can also say that the results are given in *domain* $\mathcal{R}_j$. We define a *virtual solution* $\mathcal{V}_j$ as the process of subsequently modelling $\mathcal{M}_i$ $(1 \leq i \leq j)$ in formalism $\mathcal{F}_{i-1}$, until an $\mathcal{F}_0$ model $\mathcal{M}_0$ is obtained, followed by the $\mathcal{V}_0$-solution and the subsequent enhancements $\mathcal{E}_1$ through $\mathcal{E}_j$. Schematically, the GMTF has the structure depicted in Figure 10.1. The small boxes represent system models (right-hand side) or solution results (left-hand side). The large box represents the actual numerical solution. In the context of performability modelling this will often be a Markov reward analyzer. The single-pointed arrows represent automatic translations of one formalism into another. The double-pointed arrows represent the virtual solutions.

A less general version of this figure appears in [418], where a tool for performance analysis of communication systems based on Markovian analysis is presented, and in [217] where software tools for general Markov reward models are discussed.

A similar layering structure for model-based evaluations has recently been proposed by Lepold [282] being a generalization of the open layered architecture for dependability analysis as presented in [338]. Lepold distinguishes five functional layers: a *model analysis layer* where MRMs are analytically solved; a *modelling layer* where models are described in a higher level formalism; a *problem analysis layer* where possible pre-calculations are done and where aggregation and/or decomposition strategies are chosen; the *evaluation layer* where performability problems are defined in terms of the available resources in the system to be modelled, their fault-tolerance behaviour, their workload, etc., as well as the required measures of interest; and finally the *application layer* where the design of fault-tolerant and distributed systems is the issue of interest.

Hierarchical modelling also fits in the GMTF. In hierarchical model description formalisms, high level models are described in terms of lower level models and interactions between them. This intuitive "description in terms of" conforms to the above mentioned $\mathcal{F}_{i-1}$ modelling of $\mathcal{M}_i$.

As an example of how a performance tool fits in the GMTF, consider a tool for the evaluation of generalized stochastic Petri nets (GSPNs) such

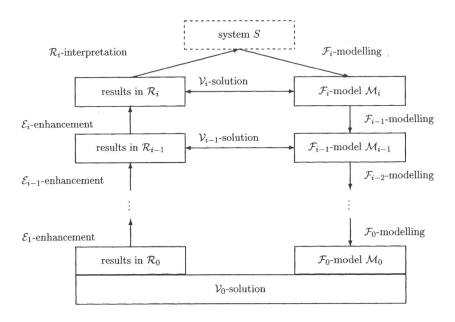

Figure 10.1: The general modelling tool framework

as GREATSPN [71]. When using this tool, a system S has to be described in the GSPN formalism. This is the $\mathcal{F}_1$ modelling activity, yielding an $\mathcal{F}_1$-model $\mathcal{M}_1$ of the system, i.e., a GSPN. This model can be converted automatically into a Markov model. This is the $\mathcal{F}_0$ modelling activity, yielding an $\mathcal{F}_0$-model $\mathcal{M}_0$ of the system, i.e., a Markov model. This model can be solved, e.g., by means of a Gauss–Seidel algorithm. This is the $\mathcal{V}_0$-solution. The results from this solution are the Markov state probabilities which fall in the domain $\mathcal{R}_0$. The system analyst who does not know that the GSPN is solved via an underlying Markov chain, cannot directly use these state probabilities. They have to be enhanced to a level which corresponds to the level at which the model was originally built. Thus the computation of probability mass functions for each place from the Markov state probabilities is an example of an $\mathcal{E}_1$ enhancement. The token distributions per place are then in the domain $\mathcal{R}_1$. Obtaining token probability mass functions for each place from the GSPN system description can be seen as a virtual $\mathcal{V}_1$-solution.

Although the GMTF can guide thinking about and development of tools, practice often tends to deviate from frameworks or reference models. As we will see, some tools allow their users to partially model their system in formalism $\mathcal{F}_1$ and partially in formalism $\mathcal{F}_0$. Sometimes the results are presented in domain $\mathcal{R}_i$ whereas the model is made in formalism $\mathcal{F}_j$ with $i < j$. In these cases one cannot speak of $\mathcal{V}_j$-solutions.

It is important to note that the levels mentioned are not absolute. A tool allowing for $\mathcal{V}_1$-solutions can from an application point of view be a higher level tool than another tool which allows for $\mathcal{V}_2$-solutions. To make this more concrete, consider the object-oriented modelling tool TANGRAM (see Section 10.4.14). As we will see later, this tool uses communicating objects for the description of Markov models. The communicating objects can be considered as models in the formalism $\mathcal{F}_1$. TANGRAM can be extended to allow for a modelling activity similar to that required when using SAVE (see Section 10.4.10). Consequently, a possible $\mathcal{V}_2$-solution of TANGRAM corresponds to the $\mathcal{V}_1$-solution of SAVE, i.e., the $\mathcal{V}_2$-solutions of TANGRAM are *not* of higher level than the $\mathcal{V}_1$-solutions of SAVE.

Regarding the transparancy of the $\mathcal{V}_0$-solutions, the following remarks can be made. In general, we can say that the $\mathcal{F}_0$-models are descriptions of Markov reward models. It can be discussed whether, e.g., the control of the choice of an algorithm in the $\mathcal{V}_0$-solutions should be made internally, or whether these choices should be part of the $\mathcal{F}_0$-models (in the form of directives). Clearly, it is very difficult to totally separate "low level details" from the "real modelling work". However, at least an intuitively appealing attempt in this direction should be made. Thus, we feel that we must be able to do higher level solutions without detailed knowledge of the underlying techniques; however, if we do know the underlying techniques, we should be able to control them.

As a conclusion of the presentation of the GMTF we would like to put forward a number of criteria related to the GMTF which should be fulfilled by a well-designed tool:

- the results should be presented in the domain $\mathcal{R}_i$ whenever the modelling is done with formalism $\mathcal{F}_i$;

- the tool should most optimally be extendible towards higher-level solutions, i.e., given that the tool allows $\mathcal{V}_i$-solutions, it should be possible to adapt the tool in such a way that it provides $\mathcal{V}_{i+1}$-solutions;

- the tool should be capable of $\mathcal{V}_i$-solutions ($i \geq 1$);

- the $\mathcal{V}_0$-solution should be transparant.

10.3 Classifying performability measures

In this section we briefly present the most important classes of performability measures. Although this section largely overlaps with Section 1.2, we include it here to keep the chapter self-contained.

Performability modelling and evaluation is applied to study the quantitative aspects (performance, dependability or performability) of gracefully degradable computer and communication systems [463]. Let S denote the set of all possible configurations in which such a system can operate. Now, define a continuous-time stochastic process $X = \{X_t, t \geq 0\}$, $X_t \in S$, describing the structure of the system at time t, i.e., which components are operational and which are down at time t. X is often referred to as the *structure-state process* since it describes the state of the system structure. The (steady-state) performance of the system when in structure state $i \in S$ is denoted by $r(i)$, where $r : S \to I\!R$ is a real-valued *reward rate function* on the state space S. Note that the function r can be determined by multiple performance analyses, i.e., for every $i \in S$ corresponding to some system configuration, the value $r(i)$, e.g., the throughput, has to be obtained by means of a "classical" performance analysis. The values $r(i)$ summarize the system performance in structure state i, instead of coping with all possible performance states in every possible structure state.

For $i \in S$, let the row vector $\underline{\pi} = [\cdots, \pi_i, \cdots]$ denote the initial probability vector on S, the row vector $\underline{p} = [\cdots, p_i, \cdots]$ the steady-state probability of residing in state i, and the row vector $\underline{p}(t) = [\cdots, p_i(t), \cdots]$ the (transient) probability of residing in state i at time t. The following measures can then be distinguished:

1. the steady-state performability (SSP): $P = \sum_{i \in S} p_i r(i)$;

2. the transient or point performability (TP): $P(t) = \sum_{i \in S} p_i(t) r(i)$;

3. the mean reward to absorption (MRTA): $\text{MRTA} = \sum_{i \in S_A} z_i r(i)$, where S_A is the subset of non-absorbing states in S and where z_i follows from a slightly adapted system of linear equations;

4. the cumulative performability (CP): $Y(t) = \int_0^t r(X_s)ds$;

5. the performability distribution (PDF): $F(t,y) = \Pr\{Y(t) \le y\}$.

The basic model discussed so far is a so-called *rate-based* reward model which means that when residing in a particular structure state $i \in \mathcal{S}$ at time t, the system performs with rate $r(i)$. The rates, however, may also depend on the global time t, thus having a reward rate function $r(i;t)$ for every state $i \in \mathcal{S}$. It is also possible to address *impulse-based* reward models. With these models a *reward impulse function* $r : \mathcal{S} \times \mathcal{S} \to \mathbb{R}$ can be defined which associates a reward $r(i,j)$ with every transition from state $i \in \mathcal{S}$ to state $j \in \mathcal{S}$. Every time a transition from state i to state j takes place, the cumulative reward increases instantaneously by $r(i,j)$ units. Of course these rewards may also depend on the global time, thus having transition reward functions $r(i,j;t)$. In general, combinations of these four possibilities may coexist. For an overview of various Markov reward models the reader is referred to Howard [236].

In the rest of this chapter we will not discuss the means to compute the above measures; this has been the topic of the preceding chapters in this book. For this chapter it is important to identify which measures can be computed with which tools.

10.4   Performability modelling tool survey

In this section we survey 17 performability modelling tools. Knowing that performability modelling incorporates both performance (for obtaining the rewards) and Markov reward model analysis, a performability modelling tool should support both of these. As we will see this is not yet commonplace. The tools we survey can, however, all in some way be used for some aspects of performability modelling. In our choice, we have restricted ourselves to Markov reward model-based tools. We do not include general simulation-based tools. Also, we do not include tools purely aiming at dependability (e.g., FIGARO [49], SURF [92]) or tools aiming merely at performance, such as GREATSPN [71], QPN-TOOL and HIQPN [23], MARCA and XMARCA [444, 443], MACOM [418], QNAUT [206, 209], or SPN2MGM [205, 207].

The following tools have been included in the survey (in alphabetical order): DSPNEXPRESS, DYQNTOOL$^+$, METAPHOR, METASAN, METFAC, NUMAS, PANDA, PENELOPE and COSTPN, PENPET, SAVE, SHARPE and GISHARPE, SPNP, SURF-2, TANGRAM-II, TIMENET, TIPPTOOL, and ULTRASAN. For each tool we:

1. provide references and implementation details;

2. discuss the model class and the supported evaluation techniques;

3. discuss the input and output format of the tool;

4. discuss how the tool fits the GMTF;

5. summarize and evaluate its most important characteristics.

It is difficult to report in general terms on the (relative and absolute) performance of the tools. The tool performance very much depends on the models one studies, i.e., their size, structure and the numerical parameters involved, the measures one is interested in (only steady-state measures or also distributions), and the actual coding of the algorithms. For a comparison of the complexity of the numerical techniques *per se*, we refer to the relevant literature discussed in the preceding chapters. How fast the various high-level model descriptions can be translated to the lower-level Markov reward models remains an interesting question. To really compare the tool performances, a number of case studies should be tackled by all the tools and the resulting performances compared. This, however, goes beyond the scope of this chapter.

10.4.1 DSPNEXPRESS

The development of DSPNEXPRESS was started by Lindemann in the early 1990s at the TU Berlin and GMD FIRST [289]. Since 1998, the development of DSPNEXPRESS has been continued at the University of Dortmund [292]. The current version has been written in C^{++} and runs on SUN (SUNOS/SOLARIS), SGI (IRIX), HP 9000 and PC (LINUX) platforms.

Model class and evaluation techniques DSPNEXPRESS allows for the specification and solution of DSPNs (deterministic and stochastic Petri nets), a class of stochastic Petri nets including immediate, exponentially timed as well as deterministically timed transitions, as originally proposed by Ajmone Marsan and Chiola [6]. However, the algorithms incorporated to evaluate these DSPNs have been improved considerably since then. In particular, DSPNEXPRESS employs a modified uniformization algorithm as well as an approach to compute Poisson probabilities in a stable manner (as proposed by Fox and Glynn [144]) [290]. The delay of the deterministic transitions may be dependent on the marking of the rest of the DSPN. In the solution process of these DSPNs, multiple independent subordinated Markov chains have to be analyzed. To speed up the overall solution process, this is done in a parallel fashion (using a cluster of workstations).

Furthermore, the restriction to have only one deterministically timed transition enabled at any time has been relaxed through new algorithms based on the method of supplementary variables [164], and on generalized semi-Markov processes [293].

With DSPNEXPRESS both steady-state and transient measures can be obtained in case only exponentially timed transitions are used. In the former case the "standard" numerical methods are included (Gauss–Seidel, SOR), whereas in the latter case randomization is the implemented method. For the DSPNs, special algorithms have been developed to compute steady-state measures, as reported in [292].

Model input and output With DSPNEXPRESS, the DSPN models are specified completely graphically. Measures of interest are specified using a flexible reward definition in terms of the DSPN, and results are displayed either textually or graphically (using the standard tool `ghostview`).

GMTF fitting DSPNEXPRESS can be used for $\mathcal{V}_1$-evaluations. Although the rewards are specified at the net level, they are not specified in a model-based way directly. This causes the tool to be less convenient for true performability modelling.

Evaluation DSPNEXPRESS is a powerful tool for the construction of Markov reward models in which not just exponential transitions are allowed but also deterministic transitions. The latter fact, including the possibility of having concurrent deterministic transitions, is the most outstanding feature of the tool. The tool provides a user-friendly graphical interface and can be used for the evaluation of steady-state and, in case no deterministic transitions are involved, transient measures.

10.4.2   DyQNTOOL⁺

The tool DyQNTOOL⁺ was designed by Haverkort [203] at the University of Twente, as a successor of DyQNTOOL [214]. DyQNTOOL⁺ has been implemented on SUN systems (UNIX), using the C programming language. For its operation, it makes use of the packages SPNP and SHARPE (see Sections 10.4.12 and 10.4.11, respectively).

Model class and evaluation techniques   DyQNTOOL⁺ is based on Markov reward models. As with DyQNTOOL, the models are specified along the lines of the dynamic queueing network concept [212, 201]. Basically, these are queueing networks that have been specified up to some parameters which are varying in time. The performance aspects of the model are described by this parametrized queueing network. The actual parameter values are a function Φ of the marking of an SPN that describes the dependability aspects. The function Φ can be very general, allowing for case recognition, etc. The generation of the Markov chain from the SPN description, the application of the function Φ and the subsequent substitution in and solution of the queueing network models proceeds completely automatically.

The class of Markov chains that can be used is the same as the class that can be used by SPNP since the SPN-part of the dynamic queueing network models are handled by that package. For the rewards, all kinds of performance measures can be used, as long as they can be derived from "SHARPE queueing networks" (or are a function of such values). The function Φ, which maps the markings of the stochastic Petri net to the parameters of the queueing network model, can be any C function.

The Markov chain derived by SPNP and the rewards derived via multiple performance analyses performed by SHARPE are combined to derive steady-state, transient and cumulative performability measures. For this purpose, the steady-state and transient probabilities and the cumulative state residence times are derived by SPNP.

Model input and output   The inputs to DyQNTOOL⁺ are a number of C files. First, there is the CSPL description of the dependability model, directly solvable by SPNP. Secondly, there is the description of the parametrized queueing network as a set of special C function calls. When this description is executed, it generates code for SHARPE. Thirdly, there is the C file describing the mapping from the SPN markings to the queueing network parameters. Finally, also by some special C function calls, the performance measures to be used as rewards are specified.

Given the above files, DYQNTOOL$^+$ generates the underlying Markov chain and the necessary SHARPE performance models. It then solves these performance models using SHARPE and combines them with the state probabilities (or residence times) derived by SPNP. It finally presents the output in terms of tables.

GMTF fitting With DYQNTOOL$^+$ the performability models are completely specified at a level higher than the Markov reward model level, i.e., as with PENPET and NUMAS, also the rewards are derived from a model. The tool therefore allows for $\mathcal{F}_1$-modelling. The final results are presented in a way that can be understood without knowledge of the underlying Markov reward model. Therefore, DYQNTOOL$^+$ is a tool that allows for $\mathcal{V}_1$-evaluations.

Evaluation DYQNTOOL$^+$ is based on the concept of dynamic queueing networks. As such, it is based on a framework in which both the dependability and the performance aspects as well as their interdependence are formally specified. The limitation to "SHARPE queueing networks" is of a practical nature and not fundamental; moreover, it might even be considered to allow for the inclusion of other modelling options of SHARPE for the derivation of rewards. The derivation of large models becomes relatively easy with DYQNTOOL$^+$; the "hand work" needed to upgrade a Markovian dependability model to a Markov reward model suited for performability analysis is done totally automatically.

Miscellaneous Recently, Das and Woodside proposed a very similar approach towards the performability evaluation of fault-tolerant systems [101]. In their approach, layered queueing networks [389] are extended to allow for the description of fault-tolerance aspects in a model. Similar to the approach followed in DYQNTOOL$^+$, multiple performance analyses are performed automatically to obtain rewards, which are subsequently used in a fault-tree look-alike method (so-called "and–or graphs") for computing performability measures.

10.4.3 METAPHOR

The tool METAPHOR (Michigan EvaluaTion Aid for PerpHORmability) has been developed by Furchtgott and Meyer at the University of Michigan [151, 152]. An APL as well as a C/UNIX version of METAPHOR exists.

Model class and evaluation techniques METAPHOR addresses only models of systems that are non-repairable and non-recoverable. Having non-repairable models implies that the Markov chain $X = \{X(t), t \geq 0\}$ is acyclic; non-recoverability means that whenever the process structure allows a transition from state i to state j then $r(j) \leq r(i)$, i.e., the modelled system cannot improve its performance over time. On the other hand, the stochastic process X may be semi-Markovian. Since the stochastic processes are acyclic, one can enumerate all state trajectories from the unique starting state to all the down states. Knowing the state transition behaviour, the probabilities of occurrence of a particular state trajectory as well as the distribution of the cumulative performance of the system over that state trajectory can be computed. Combining these two yields the performability distribution $F(t, y)$. In METAPHOR, techniques have been implemented to reduce the number of state trajectories. No explicit mention is made of the maximum size of models METAPHOR can cope with. However, an example with hundreds of state trajectories is discussed by Furchtgott [152].

Model input and output The models are input interactively. The reward rates for all possible structure states have to be input separately. This implies that a separate performance modelling activity is necessary. The output of the tool gives values for the function $F(t, y)$ in tabular form.

GMTF fitting METAPHOR is an $\mathcal{F}_0$-modelling tool. The output is in the $\mathcal{R}_0$-domain. The tool can be used for $\mathcal{V}_0$-evaluations.

Evaluation METAPHOR was the first tool for performability modelling. Its model class is restricted in the sense that only acyclic (non-repairable), non-recoverable stochastic processes can be used. There are, however, no restrictions on the state residence distributions. As models become large, the enumeration of all the state trajectories and the integration over them can be a severe drawback in using the tool. Another disadvantage of the tool is the fact that the possibly large number of reward rates has to be calculated off-line. Both the input and the output of the tool are textual.

10.4.4 METASAN

The tool METASAN (Michigan Evaluation Tool for the Analysis of Stochastic Activity Networks) was developed by Meyer and Sanders and colleagues *et al.* [318, 410, 411, 405] at ITI (Industrial Technology Institute) in cooperation with the University of Michigan. It was written in C with the help of the tools Yacc and Lex. Versions for SUN 3 and VAX (both UNIX) exist.

Model class and evaluation techniques The tool METASAN is based on the Stochastic Activity Network (SAN) approach of describing complex systems [318]. SANs consist of places and activities (transitions in GSPN terminology). Activities can be either stochastically timed or instantaneous. In the timed case, general distributions are allowed; however, if non-exponential distributions are used, simulation should be used to solve the model. So-called cases can be associated with each activity. Cases are generalizations of probabilistic switches [28]. Upon completion of an activity, one of the cases is chosen probabilistically. Both the timing distributions and the case probabilities can be marking dependent. Input and output gates that connect places to activities and *vice versa* determine the flow of tokens through the SAN. Input gates have a predicate that specifies whether the activity is enabled or not, and a function that specifies how the tokens from the input places are redistributed over the output places of the activity. Output gates only have the latter function; they do not have predicates.

With METASAN, it is possible to construct models in a hierarchical fashion using a macro facility, i.e., submodels, SANs in itself, can be defined which are subsequently used (one or more times) in higher level models. The overall SAN model describes the structure-state process. In terms of place occupancies, the user can associate reward rates with all states. From these, an underlying Markov reward model is derived. The underlying Markov model is solved for the steady-state performability measures using either direct or iterative methods. Transient performability measures are obtained using uniformization. A number of algorithms have been implemented for obtaining $F(t, y)$ under various model restrictions. A specialized algorithm is incorporated for obtaining $E[Y(t)]$.

It is also possible to use simulation as an evaluation technique. Steady-state as well as transient measures (terminating simulation) may be obtained in this way. Confidence intervals are also obtained and can be used to specify stop criteria for the simulation, e.g., stop if the 95% confidence interval width is within 10% of the estimate.

METASAN is one of the two performability modelling tools (the other one is ULTRASAN; see Section 10.4.17) that combines rate-based and impulse-based Markov models.

Model input and output Model input and output are textual. The input consists of two parts: a part describing the actual model and a part describing the experiment, i.e., the measures to be obtained, the techniques to be used, etc. The model description part is done with the SAN description language SANSCRIPT and is independent of the experiment description. Output is given in tables.

GMTF fitting With the METASAN modelling language, the modelling activities take place at a higher level than at the Markov reward level, i.e., $\mathcal{F}_1$-modelling. Due to the macro facilities, a form of higher-level modelling is also possible. Reward rates are specified as a general C function of the SAN marking and/or the activity completions. As such, the reward rates are not derived from a performance model, albeit that the C functions can be an "implementation" of some queueing model. The output is given in a form that can be understood at the level of model description, i.e., in the $\mathcal{R}_i$-domain ($i \geq 1$). Thus, $\mathcal{V}_i$-evaluations ($i \geq 1$) can be done with the tool.

Evaluation METASAN provides a very general framework for building performability models. The tool supports analytical as well as simulative evaluation techniques. The separation between a model description and a model evaluation part is a nice feature since it allows the modeller to evaluate one particular model with various evaluation techniques. Input of the tool is textual. Output is in the form of tables. Reward rates have to be computed separately.

10.4.5 METFAC

The tool METFAC (Modelación y Evaluación de la Tolerancia a Fallos Asistida por Computador) has been developed by Carrasco and Figueras at the Polytechnical University of Catalunya [61, 60]. It has been written in Fortran 77 and runs on VAX (VMS) systems.

Model class and evaluation techniques The METFAC system allows the analysis of Markov reward models. The models are specified by means of a production rule system [217, Section 4.4]. The system structure is described by a number of state variables. For all possible values (value ranges) of the state variables, the events that can occur, the rate at which they occur and how they affect the state variables, are specified. In this way, all possible finite Markov chains can be described. Steady-state as well as transient performability measures can be obtained. A measure close to the cumulative performability measure can also be obtained: the so-called serviceability $S(y)$, i.e., the probability distribution that an amount of work y has been done before the first failure occurred. For obtaining steady-state measures direct methods are employed. The transient measures are obtained using numerical integration procedures especially suited for stiff problems. Large state spaces are dealt with by an approximate technique called dissolving. Dissolving basically means that states with very low probabilities are omitted but accounted for by altering transition rates in their predecessor states. The evaluation routines in METFAC are especially developed for large, sparse Markov models. It is suggested that the tool is able to deal with models having thousands of states.

Model input and output The production rule system has to be input textually. The reward rates have to be input separately for all possible states. The translation from the production rule system to the underlying Markov reward model is done automatically. Output is given in tabular form.

GMTF fitting Modelling systems with the production rule mechanism of METFAC typically is an $\mathcal{F}_1$-modelling activity. The specification of the reward rates to be associated with every global state is a lower level model activity, i.e., $\mathcal{F}_0$-modelling. Output of the tool is in the $\mathcal{R}_0$-domain.

Evaluation METFAC provides a very powerful mechanism for constructing the dependability part of a Markov reward model that can subsequently be used for performability analysis. Reward rates have to be supplied manually. A large variety but not all of the performability measures can be obtained. In the implementation of the tool, much emphasis has been put on state dissolving techniques and on the exploitation of the sparsity of the infinitesimal generator matrix in the evaluation routines.

10.4.6 NUMAS

NUMAS (NUmerical Methods for the Analysis of computer Systems) has been developed by Müller-Clostermann at the University of Dortmund [336]. NUMAS is basically a performance analysis tool; however, it is possible to extend each queueing station with fault-tolerance characteristics, so that performability measures can be obtained. NUMAS has been implemented in Simula 67 for Siemens BS2000 systems. The algorithms used in NUMAS are also accessible from the hierarchical modelling tool HIT as discussed by Beilner, Mäter and Weissenberg [30]. For HIT, versions exist for IBM, APOLLO and SUN systems. A graphical interface is also available.

Model class and evaluation techniques With NUMAS queueing network models are solved by numerically computing the steady-state distribution of the underlying Markov chain. Within every multi-server queue individual servers may fail with load-dependent rates and can be repaired at a rate dependent on the number of servers that is still available, independently of all other queueing stations. Thus, the state of the network consists of a part describing the distribution of customers over the queueing stations and a part describing the so-called degradation mode of each queueing station. Groups of states with the same degradation mode are said to belong to one macro state. By the fact that there are large timescale differences between transitions inside and between macro states, it is possible to decompose the overall model. For every macro state, a Markovian submodel for the associated performance model is identified and analyzed. The submodel results are used in a numerical analysis of the Markov chain describing the transitions between degradation modes.

For obtaining the steady-state probability vectors iterative and direct methods are used such as Gaussian elimination, Gauss–Seidel iterations and iterative aggregation. Models with thousands of states can easily be handled.

Model input and output The methods in NUMAS can be accessed via two user interfaces. Both allow the user to input the so-called degradable queueing network models at a high level. The associated Markov model is generated automatically. There is an interactive NUMAS interface available in the Siemens/Simula implementation. The NUMAS methods can also be accessed via the hierarchical modelling tool HIT. The output of the tool is given in tabular form and is related to the original modelling constructs.

GMTF fitting NUMAS input takes place via a specialized interactive dialogue and its output is given in terms of the input modelling elements. The user is not aware of the underlying Markov reward models. Consequently, NUMAS allows $\mathcal{V}_1$-evaluations.

When using the HIT system and HI-SLANG (HIT System LANGuage) as input language, there are very nice features for hierarchical modelling. The outputs can then be tailored to the highest level of the model description. With HIT, we thus have $\mathcal{V}_i$-evaluation capabilities ($i \geq 1$).

Evaluation For steady-state performability analysis, NUMAS provides powerful constructs. All necessary information is computed by the tool. A limitation of the tool is that it is not possible to model dependencies in the failure and repair processes of the various queueing stations. Also, only steady-state measures are computed.

10.4.7 PANDA

The tool PANDA (Petri Net Analysis and Design Assistant) has been developed by Horton and Allmaier at the University of Erlangen-Nürnberg [9] and is available for several versions of the UNIX operating system. For its graphical user-interface, the X11 and Motif libraries have been used.

Model class and evaluation techniques The modelling paradigm followed with PANDA is that of SPNs with exponentially distributed firing times, including the well-known extensions for inhibitor arcs, enabling functions, etc., as known from the package SPNP. Indeed, the input language for PANDA largely conforms to that of SPNP, i.e., CSPL. To simplify the modelling process, transition distributions can directly be specified as Erlang, hyper- and hypoexponential and Cox distributions. When these distributions are employed, different memory policies can be specified: resampling, age memory and enabling memory. Non-exponential timing can also be specified, e.g., deterministic, Normal, etc., but then only a discrete-event simulation solution is supported. In case the SPN can be translated to a finite CTMC, steady-state measures are computed with the very efficient multi-level method [235]. Transient measures are computed via uniformization. In addition to standard results (place occupancies, transition throughputs, etc.) more complex measures can be defined using rewards.

To ease the specification of dependability models, a fault-tree based input format can also be used; it is automatically transformed to an equivalent SPN before the actual solution takes place.

The most outstanding feature of PANDA is that it allows for a distributed generation and solution of the CTMC, using either a cluster of workstations (distributed memory) or a dedicated shared-memory multiprocessor [10].

Model input and output In addition to the textual input format, a graphical user-interface is provided, allowing for the specification and animation of the SPN. The GUI supports hierarchical modelling, however, the analysis is always performed on the complete "flat" model.

GMTF fitting PANDA can be used for $\mathcal{V}_1$-evaluations using SPNs. Reward rates are input at the SPN level as a general C function over the markings; the derivation of the reward rates is not model based. When using the fault-tree input format, a higher level of evaluation ($\mathcal{V}_2$) becomes possible.

Evaluation PANDA comprises a flexible tool for SPN evaluation. Among its innovative features are the use of the multilevel steady-state solution method and the use of parallel and distributed computing techniques to speed up the model solution.

10.4.8   PENELOPE **and COSTPN**

PENELOPE (dependability evaluation and optimization of performability) has been developed by de Meer and colleagues, initially at the University of Erlangen-Nürnberg and later at the University of Hamburg [304, 307, 309, 308], as a tool for performability optimization. More recently, they developed the so-called "controlled stochastic Petri nets" (COSTPN) [305, 306] for the optimization of SPNs; tool support for them will be integrated in PENELOPE in the near future. PENELOPE has been developed in C for UNIX systems, thereby making use of the X11 and Motif windowing systems, as well as of the mSQL database.

Model class and evaluation techniques The aim of PENELOPE is to allow users to optimize systems with respect to their performability, either in the steady-state or in the transient case. For this purpose, the authors have developed extended Markov-reward models (EMRMs) [304, 309], combining Markov-reward models with Markov decision processes [460]. In this approach, normal MRMs can be described, though, with the addition of reconfiguration arcs. These arcs allow for an instanteneous transition between states, and they can be controlled for optimization purposes. Similarly, branching states can be used as timeless states in which impulse rewards are gained. A *strategy* then is the description of the actual instantiation/use of reconfiguration arcs or branching states, where a strategy might be time-dependent. An optimal strategy then is a strategy which yields the largest value for the defined measure of interest, over all possible strategies. Basically, three measures of interest can be optimized: the steady-state reward rate, the expected accumulated reward up to time t, and the expected accumulated reward up to absorption (for models with absorbing states).

Typical optimization examples include the question whether or not to reconfigure a fault-tolerant system after a failure has occurred, or whether or not to accept more connections in a communication system. The best decision to be taken depends on the mission time left, the relative gain in performance after reconfiguring (or after accepting the new connection), and the costs directly associated with the reconfiguration. It is important to note that reconfiguration decisions strongly depend on the mission time left, hence transient optimizations normally yield substantially different strategies than do steady-state optimizations.

The transient optimization methods supported apply for both discrete-time and continuous-time models and form an extension of [277]. The steady-state optimization methods are based on value-iteration methods [460].

Apart from strategy optimization, it is also possible to evaluate performability models under arbitrary specified fixed strategies, using numerical solution techniques as well as simulation.

With COSTPN, de Meer recently eased the EMRM specification significantly by using SPNs for this purposes. In the same vein as for MRMs, these SPNs are extended with reconfiguration transitions.

Model input and output Using PENELOPE, models are specified at the state level using a graphical user-interface. Alternatively, textual specifications are possible. Separate modules exist for experimentation specification, result analysis and strategy presentation. The reward rates need to be assigned directly to the states. A database module is provided for handling multiple models and for allowing model reuse. With the controlled

SPNs, the specification will be completely performed at the SPN level.

GMTF fitting The state-based specification of the EMRMs in PENE-LOPE should be regarded as an $\mathcal{F}_0$-modelling activity, although one might argue that the possibility of performing an optimization over such a model forms a higher-level modelling activity. After the complete integration of COSTPNs in the tool, $\mathcal{F}_1$-modelling is possible, although one might also argue here that the optimization process over such models forms an $\mathcal{F}_2$-modelling activity. In both cases the reward rates have to be specified directly to states (markings).

Evaluation PENELOPE is the first performability evaluation tool to fully support the optimization of performance measures of interest, in both steady-state and transient, by specifying optional reconfiguration possibilities. As such, this tool provides system engineers and on-line system management entities with the proper means to make decisions. The fact that the model specification is still state-based is a disadvantage; however, we expect this to change soon when the integration of COSTPNs in PENELOPE has been completed.

10.4.9 PENPET

The tool PENPET (PEtri Net based Performability Evaluation Tool) has been developed by Lepold *et al.* at Siemens AG, in cooperation with the University of Mulhouse [280, 281, 282]. The system has been built with the C programming language under the UNIX operating system, making use of X-WINDOWS and MOTIF.

Model class and evaluation techniques With PENPET the dependability model part is at the highest level described by a structure formula. Models (so-called macro-molecules) are defined that consist of a replication of a number of submodel parts (molecular clusters) and/or components (atoms). This definition is done by means of a so-called system structure formula, a textual description that closely resembles the description of molecules in chemistry. For each of these submodel parts and components, how many there are and how many need to be operational for the next hierarchically higher model to be operational are specified. Failure rates and coverage factors are also specified per component and/or per submodel.

Components and submodels used in this way are described by one of a number of library-provided GSPNs. Extensions to this library of standard submodels can also be made. The class of GSPNs is the so-called Performability Adapted SPNs (PASPNs), which fall in the class of GSPN models that can be analyzed with TOMPSIN, a package also developed at Siemens [279]. A nice feature of TOMPSIN is that it allows for approximate hierarchical modelling, which decreases the size of the state space of the performance models [265].

From the system structure formula an overall GSPN describing the dependability aspects is derived. From this GSPN, all the possible system structure states are derived as well as the Markov chain underlying the GSPN. From all the system structure states, via a set of functions (as in the case of dynamic queueing networks, see Section 10.4.2), the parameters of the performance models are derived. These parameters might include initial markings, firing rates, etc. A reverse dependency is also possible, e.g., failure rates in the structure state model might be dependent on the reward rates that are derived; this then requires an iterative solution process. How this is exactly organized, and whether such an iteration always converges, does not become clear from the available literature. Once the performance models have been constructed they are evaluated by doing normal GSPN analyses, using the package TOMPSIN. Combining the structure state process with the obtained reward rates, a Markov reward model is obtained.

For deriving steady-state performability measures such algorithms as Gaussian elimination, Gauss–Seidel iterations, SOR and the LSQR algorithm (an extension of the conjugate gradient method) have been implemented. For deriving point performability measures the ACE algorithm [300] has been implemented as well as the randomization method.

Model input and output The input to PENPET is textual. The user can textually input the system structure formula as well as the needed GSPN submodels, if they are not readily available form the library. The output of the tool is in the from of tables and graphs.

GMTF fitting The tool PENPET allows for the construction of Markov reward models by techniques that are specially designed for the performability analysis of fault-tolerant multiprocessor systems. Also the reward rates are derived automatically from a high-level SPN model, therefore

$\mathcal{F}_1$-modelling is possible. The output is presented in the $\mathcal{R}_1$-domain, so PENPET is capable of $\mathcal{V}_1$-evaluations. The structure state formulae can be seen as $\mathcal{F}_2$-models since they are higher-level descriptions of GSPNs.

Evaluation PENPET allows for the construction of Markov reward models by techniques that are especially designed for the performability analysis of fault-tolerant multiprocessor systems. The way of describing higher-level models as clusters of lower-level models that can be specified separately is very flexible. Also the mutual dependence between performance models and structure state models is a very nice feature.

10.4.10   SAVE

SAVE (System AVailability Estimator) has been developed by Goyal *et al.* [180, 183, 182] at IBM Yorktown Heights in cooperation with Duke University. It has been written in Fortran 77 and runs on an IBM System/370. Originally designed for modelling ultra-dependable computer systems, it has been extended for performability modelling.

Model class and evaluation techniques SAVE allows for Markovian structure state processes. With every possible structure state a performance level can be associated. Steady-state measures are obtained using the iterative technique of successive over-relaxation (SOR). Transient measures are obtained using the randomization approach. Various specialized algorithms have been implemented to compute $F(t, y)$. A special feature is the possibility of obtaining the sensitivity of the results with respect to model parameters [181, 219]. Simulation of Markovian models, which is imperative when the models become very large, is also possible. In that case an importance sampling technique is used, which greatly reduces the required simulation time [185] (see also Chapter 8 of this volume). As far as the Markov approach is concerned, models with hundreds of thousands of states can be handled.

Model input and output The user can choose the Markov input approach or the higher-level input approach. In the former, the user has to specify all the Markov states, their corresponding rewards and their outgoing transitions. Alternatively, the user can specify the models at a higher level, using powerful basic modelling constructs. In both cases, the rewards have to be supplied by the user. Given a model input, SAVE automatically constructs the corresponding Markov reward model after which the user can interactively indicate which measures have to be computed. All input is textual. All output is given in tabular form.

GMTF fitting The SAVE input language is especially developed for modelling the dependability of systems, and thus supports $\mathcal{F}_1$-modelling activities. The input of the performance aspects, i.e., the rewards, is a typical $\mathcal{F}_0$-modelling actvity. The output is given in terms of the modelling constructs that are used when specifying the dependability aspects.

Evaluation SAVE is a powerful tool for the analysis of Markov reward models. Its constructs for specifying the structure state process are very powerful and so are the implemented evaluation techniques. Steady-state, transient and cumulative performability measures can be computed.

10.4.11 SHARPE and GISHARPE

SHARPE (Symbolic Hierarchical Automated Reliability/Performance Evaluator) has been developed by Trivedi and colleagues at Duke University [399, 402, 401, 400, 404]. SHARPE allows for various types of modelling and for various types of evaluation techniques. In this section we will discuss the capabilities of SHARPE only as far as performability is concerned. SHARPE has been implemented in ANSI C and runs on any platform. The graphical user-interface is in JAVA and is available on WINDOWS and SOLARIS platforms.

Model class and evaluation techniques SHARPE allows for the construction and analysis of Markov and semi-Markov chains. The Markov chains must be either acyclic, irreducible (every state is reachable from every other state) or phase-type (there is at least one absorbing state and every non-absorbing state is transient). The semi-Markov chains must be acyclic or irreducible. With every state a reward rate can be associated.

Assuming the system to be down in the absorbing states and operational in the transient states, the phase-type Markov chains can be used to calculate the system reliability, the mean time to failure, and the time to failure distribution symbolically. By using convenient reward rates, measures such as mean computation before failure (originally defined by [27]), the distribution of the cumulative reward upon absorption, the expected reward rate at time t and the expected cumulative reward at time t can be obtained. Reward rates can be computed from another model, which can be a Markov chain, a GSPN, a semi-Markov process, a product-form queueing network or a directed acyclic task graph.

When the (semi-)Markov chain is irreducible, the steady-state performability can be calculated. When acyclic semi-Markov chains are used, not only exponential distributions but also exponential polynomial distributions ("exponomials") can be used, i.e., distributions of the form $F(t) = \sum_i a_i t^{k_i} e^{b_i t}$.

The steady-state results are obtained either with Gauss–Seidel iterations or by using SOR. Transient quantities are derived in semi-symbolic form [399, 402, 401, 400, 403, 404, 382] or in numeric form by using randomization [192, 198, 104, 106].

Model input and output The specification of the Markov models is performed by enumerating all the states, the state transitions and the reward rates associated with the states. A nice feature of SHARPE is that it is possible to model hierarchically, also between different types of models. For instance, it is possible to have as a submodel in a fault-tree (also supported by SHARPE) a Markov chain model. Similarly, the reward rates can be generated automatically from a lower-level performance model.

WWW-based input and output: GISharpe Recently, Puliafito *et al.* developed a world wide web (WWW) based graphical user-interface for SHARPE, called GISHARPE [376]. With this interface, the modeller can specify his models using an ordinary WWW-browser such as Netscape, thereby using local computing resources. The actual solution of the models is performed with the SHARPE package running on the server providing the WWW interface. The advantage of such a solution is that the (central) provisioning entity of the computational routines, i.e., the server providing the interface and the SHARPE package, can easily be updated and extended, without the need to have reinstallations at the user sites. On the other hand, the server providing this service cannot be expected to perform this task free of charge. Therefore, in the future such WWW-based interfaces will most probably be extended with mechanisms for user registration and tariffing. Although such measures would bother the end-user of the tool slightly, we think that WWW-enabled graphical user-interfaces have a great potential, as also claimed by van Moorsel and Huang [331].

GMTF fitting By computing reward rates using a lower-level model such as a queueing network, SHARPE allows one to perform $\mathcal{V}_1$-evaluations in a similar way to DYQNTOOL$^+$.

Evaluation The real strength of SHARPE lies not in pure performability modelling, but more in its capability to use various modelling techniques in a combined fashion. The input formalism of SHARPE has recently been made more attractive for performability evaluation purposes, e.g., by the so-called loop facility. Furthermore, the reward rates can be computed from a model. The recent extension to GISHARPE greatly contributes to the usability of SHARPE. Similarly, recently developed graphical user-interfaces have increased its ease of use.

10.4.12   SPNP

The Stochastic Petri Net Package has been developed by Ciardo, Muppala and Trivedi at Duke University [82, 83]. SPNP is a very general and flexible package for the construction of stochastic Petri nets (SPNs) and some extensions, such as non-Markovian SPNs and fluid SPNs. The class of SPNs supported by SPNP is sometimes referred to as stochastic reward nets (SRNs). The underlying Markov reward models are solved using analytic-numerical methods (if possible) or simulatively otherwise. SPNP has been implemented in C and runs on SUN, CONVEX and NEXT systems (all UNIX), RS/6000 (AIX), PS/2 (OS/2) and WINDOWS.

Model class and evaluation techniques With SPNP all types of Markov reward models can be constructed. There are no limitations on the reward rates. The Markov reward models are derived from an SPN description. The reward rates can be specified as normal C functions over the number of tokens in the places. This allows for a very flexible reward structure construction, since the C functions used can invoke other functions, e.g., to do some performance analysis.

In fluid SPN, continuous variables can be introduced through fluid places, which enable the modelling and analysis of more complex systems. Likewise, the possibility of using non-exponential distributed transition firing times removes a major restriction of many SPN tools.

The resulting Markov reward models can be analyzed for steady-state measures, using Gauss–Seidel iterations, SOR or the Power method. For Markov reward models with absorbing states, the cumulative reward until absorption can be calculated. Transient and cumulative measures are obtained using uniformization. Sensitivities can be obtained for steady-state measures, again using Gauss–Seidel, SOR or the Power method, and for transient measures, using uniformization.

In version 6 of SPNP, the modelling capacity has been extended by including non-Markovian SPN (almost 20 different distributions are implemented), various resampling policies and fluid SPN. Facilities are provided to perform series of experiments, to solve models hierarchically, and to apply fixed-point iterations over (sub-)models. To analyze non-Markovian SPNs, fluid SPNs and very large ordinary SPNs, discrete-event simulation is also supported. Several simulation methods have been implemented: independent replicas, batch means and regenerative simulation. Speed-up techniques such as importance sampling and importance splitting are also supported.

Model input and output The input language for SPNP is CSPL, i.e., the C-based SPN Language. A CSPL description is a normal C file; it is compiled with the C compiler and linked with a number of other C programs which together constitute the SPNP. The output of the tool takes place via C functions. This allows the tool user to tailor the output to his specific needs.

Recently, the integrated SPN modelling environment has been developed, which provides a graphical user-interface to SPNP [231].

GMTF fitting SPNP can be used for $\mathcal{V}_1$-evaluations. The possibilities of the tool allowing for the use of parametrized subnet specifications indicate that the modelling and evaluation can be enhanced to higher levels than the standard $\mathcal{F}_1$-modelling and $\mathcal{V}_1$-evaluations. Reward rates are input at the SPN level as a general C function over the markings. Note that through this possibility, reward rates can be derived in a model-based fashion.

Evaluation SPNP is a powerful tool for the construction of Markov reward models. Since the model description is done via standard C, the full power of the C programming language is available. This makes the tool very flexible and easy to use. This, as well as the fact that rewards can be associated with every marking by using normal C functions, opens up numerous possibilities to further facilitate the modelling process. A wide variety of performability measures can be obtained, including steady-state, transient and cumulative measures.

10.4.13 SURF-2

SURF-2 has been developed at LAAS-CNRS as a high-level tool for dependability and performability evaluation [31] as a successor of SURF [92]. It has been implemented for SUN systems running UNIX, using the languages C and ADA.

Model class and evaluation techniques Models can be either Markov reward models or GSPNs. The latter models are automatically transformed to the former before the analysis is started. Steady-state measures, transient measures and MRTA measures can be derived by the tool. Which techniques are implemented is unclear from [31].

Model input and output Both the Markov reward models and the GSPNs are input graphically. Output can be represented using a so-called result formatter which allows for the graphical representation of the results. As with ULTRASAN (see Section 10.4.17), in the model definitions one can refer to global variables. By specifying the range of these variables, one can easily do series of analyses which, in combination with the result formatter, allows for the easy derivation of graphs.

GMTF fitting SURF-2 allows for $\mathcal{V}_0$-evaluations when the Markov reward model input is chosen and for $\mathcal{V}_1$-evaluations when the GSPN input option is chosen. Reward rates are specified as a general function over the markings, i.e., not model based. In total, $\mathcal{V}_1$ evaluations are possible.

Evaluation SURF-2 is a powerful tool for the execution of series of dependability and performability analyses. Based on the powerful GSPN concept every required model can be made. Graphical input and output ease the modelling task. A limited set of measures can be derived.

10.4.14 TANGRAM-II

TANGRAM is a general object-oriented tool environment that has been developed in cooperation by the University of California at Los Angeles and the Federal University of Rio de Janeiro [34, 57, 58, 368, 106]. It provides a layered tool environment which can very easily be tailored to specific application domains. In what follows, we address the latest version, i.e., TANGRAM-II [57, 58], which has been implemented in C and C++.

Model class and evaluation techniques TANGRAM-II can be used for Markov-reward based performance, dependability and performability analysis. Its most distinguishing property is its extremely flexible model description format, based on object-orientation. Models are specified as collections of objects. Objects are parametrized instances of object types. Every object type has an internal state which changes upon the completion of internal events (internal states can be seen as vectors of integers). Internal events can also cause messages to be sent to other object types. Reception of messages by an object causes events within the receiving object. The overall model state is the collection of the internal states of all the objects. With every state a reward rate can be associated. Since object types can inherit properties from parent object types, complex models can be constructed in a stepwise fashion. In this way it is possible to define a set of high-level objects specially tailored for a particular application.

TANGRAM-II supports the evaluation of the Markov-reward model underlying the object-oriented model specification with a wide variety of techniques. As direct methods for steady-state evaluations, GTH and block GTH (a (blocked) variant of Gaussian elimination that avoids subtractions [195]) and a new efficient algorithm are included [312]. As iterative methods, SOR, Gauss–Seidel, Jacobi and Power are included. Transient performability measures (probabilities, expected rewards, cumulative rewards and distributions of the latter) are computed with uniformization. For non-Markovian models, a new solution method for steady-state analysis has been incorporated [112]. Finally, discrete-event simulation is provided as well.

For each object, reward rates can be associated with states, provided certain conditions are fulfilled. Reward rates are then defined by combining these "per-object" rates.

Model input and output The input of TANGRAM-II consists of a mixture of textual and graphical components, thereby making use of the TANGRAM graphic interface facility TGIF. The output of TANGRAM-II is given in textual form and currently mainly addresses probabilities and simple rewards.

GMTF fitting The TANGRAM-II tool environment allows for the easy construction of Markov-reward models. Models can be built hierarchically, so we typically have to do with $\mathcal{F}_i$ ($i \geq 1$) modelling. The definition of reward rates is not model-based. TANGRAM-II typically allows for $\mathcal{V}_i$-evaluations ($i \geq 1$).

Evaluation TANGRAM-II is a flexible modelling environment to construct Markov-reward models. It allows for hierarchical modelling and supports a wide variety of evaluation methods.

10.4.15 TimeNET

The tool TIMENET has been designed by German *et al.* [163] at the Technical University of Berlin, as a successor of DSPNEXPRESS (see Section 10.4.1). TIMENET has been developed in C and C++ and runs on SUN (SOLARIS 2) and PC (LINUX) platforms. Its graphical user-interface uses the X11 window environment.

Model class and evaluation techniques TIMENET supports the modelling with and evaluation of a large class of stochastic Petri net models: normal SPNs, including exponentially timed and immediate transitions [5]; DSPNs, including the enabling of at most one deterministic transition in any marking (in addition to exponential ones) [6]; extended DSPNs, which allow for the enabling of at most one exponentially-polynomially distributed transition (next to the normal exponential ones) [78]; and concurrent DSPNs, which allow for the enabling of concurrently enabled deterministic transitions (in addition to the usual exponential ones) [164].

The SPN models are specified completely graphically. Special popup menus are provided to specify the exponentially-polynomial distributions. To validate the constructed models, an interactive token game can be played. Furthermore, place invariants and extended conflict sets of immediate transitions are also computed [25].

Reward rates are specified in a similar style as in DSPNEXPRESS and, earlier, in GREATSPN, using simple expressions over the markings in the SPN, for which individual probabilities or expectations can then be computed.

The tool supports the computation of steady-state and transient measures for extended DSPNs (including the class of DSPNs and normal SPNs), thereby using methods based on uniformization, supplementary variables and Runge–Kutta techniques, as well as a variety of direct and iterative linear system solvers. Furthermore, an approximation procedure for steady-state measures in the case of concurrent DSPNs (based on generalized Cox distributions) has been developed [158, 165]. In addition to these methods, TIMeNET supports the simulation of general SPNs (irrespective of the employed distributions) for both steady-state and transient measures. In that case, variance reduction methods based on control variates, as well as on the RESTART method (especially suited for the estimation of rare events), are supported [261].

As a complement to continuous-time models, also discrete-time models can be specified and evaluated, in both stationary and transient cases [480].

Model input and output TIMeNET has a graphical user-interface to specify the models. The tool also supports various output formats, including simple tables, as well as various graphs; the latter can even display the results of transient analysis as they are computed. There is also support for the definition of multiple-run experiments. Recently, two extensions of TIMeNET have been completed:

(i) TIMeNET-MS [479], which is especially tailored towards manufacturing systems for which purpose the modelling formalism has been extended by coloured tokens (among others) and which provides a steady-state analysis and a simulation component;

(ii) TIMeNET-SPNL [160] which represents an approach to structured modelling and which combines graphical Petri net primitives with structural programming language constructs. At the moment models are mapped on SPNs and can be simulated. Therefore, TimeNET-SPNL can be considered as an E_2-enhancement.

GMTF fitting TIMeNET can be used for $\mathcal{V}_1$-evaluations. Reward rates are input at the SPN level using simple functions (additions and comparisons) over the markings. As with many other tools, this derivation is not model based. TIMeNET-SPNL can be regarded as a tool allowing for $\mathcal{V}_2$-evaluations.

Evaluation TIMeNET is a very powerful tool for the evaluation of a wide variety of (non-exponential) SPNs, including GSPNs, extended DSPNs, concurrent DSPNs and discrete-time SPNs. Steady-state and transient measures are supported. Discrete-event simulation is supported as an alternative.

10.4.16 TIPPTOOL

TIPPTOOL [224] is a representative of a class of new tools based on stochastic process algebras (SPAs) [225]. Other tools in this class are the PEPA workbench [168] based on the stochastic process algebra PEPA [230] and EMPA [33, 32] (now called TwoTowers).

TIPPTOOL has been developed at the University of Erlangen-Nürnberg for UNIX-based systems, using the language C for the numerical algorithms, and Standard ML for the modelling language parser and state-space generator (this language is very well suited for implementing semantics of formal languages; also the PEPA workbench has largely been implemented with it).

Model class and evaluation techniques With SPAs, stochastic models are constructed in a component-wise action-oriented manner. Model components are specified using stochastic extensions of "classical" process algebra, in which timing delays are associated with some actions, and where the other (untimed) actions are assumed to be of immediate type.[1] In most SPA-based tools, as in TIPPTOOL, the delays are represented as exponentially distributed random variables.

Given a number of model components, they can be combined and synchronized with one another, to end up with a model of a larger subsystem or with a complete system model. This compositional way of describing models is one of the key assets of SPA-based modelling approaches; it provides a very natural means for hierarchical model construction.

Given a complete SPA-based model, the underlying CTMC can be generated automatically, thereby using the so-called structural operational semantic rules. In practice, an important feature is the fact that aggregations can be employed per component, that is, under certain restrictions summarized in the notions of Markovian bisimularity and weak Markovian bisimularity which are indeed closely related to lumpability in Markov chains [55], individual components can be minimized (made smaller in terms of the number of states and transitions) without affecting the measures of interest to be computed from an overall model in which these minimized components are embedded (substitution property). For details, refer to

---

[1] In some SPAs, most notably in Hermanns' interactive Markov chains [223], actions and delays are explicitly separated, in order to avoid the problem of expressing the delay of synchronized timed actions.

[224] or the example case study [226]. In TIPPTOOL efficient minimization methods based on so-called partial refinement have been implemented [227].

The association of reward rates to states is less natural in SPA-based approaches, as these approaches have only an implicit notion of system state; the notion of state-change (action) is most prominent. Nevertheless, approaches do exist to associate reward rates to states, e.g., via an association to actions leading towards particular states [32], or via the use of modal logics to select groups of states that are to be coupled to particular reward rates [89].

With TIPPTOOL the notion of reward rates is not explicitly available. Using standard direct and iterative methods, steady-state probability vectors can be computed, as well as transient probability vectors, using uniformization. Three types of measures are supported: (i) state measures allow one to compute the probability (steady-state or transient) of groups of states, e.g., to compute availability or resource utilization; (ii) throughput measures allow one to compute the occurrence frequency of particular actions; and (iii) mean-value type measures allow one to compute mean values for particularly structured submodels, e.g., to compute mean queue lengths.

In addition to exact methods, approximate methods based on near-complete decomposability (time-scale decomposition) and response time approximations are also supported. For details, refer to [313].

Model input and output The SPA models are specified textually, although in a parametric fashion. Experiments can be defined separately, in order to study a system under varying parametric conditions (the underlying CTMC is generated only once). Numerical results can be output textually, or can be exported to be displayed with the PXGRAPH package.

GMTF fitting With TIPPTOOL model specifications are completely specified, in a hierarchical fashion, at the SPA level. As far as the system dynamics are concerned, the tool allows for $\mathcal{F}_i$-modelling ($i \geq 1$). However, reward rates cannot be explicitly and generally assigned. Since only standard measures can be computed anyway, this imbalance is not directly visible. Nevertheless, there is a need for a more flexible and powerful way to express more general reward-based measures.

Evaluation TIPPTOOL as a representative tool based on SPAs provides very powerful means to specify complex models of systems in a hierarchical and compositional way. The potential of compositional modelling and its exploitation via component-wise aggregation in the solution process should not be underestimated. For these reasons, SPA-based tools have the potential to become more important in the future.

10.4.17 ULTRASAN

The tool ULTRASAN has been designed by Sanders *et al.* [96] at the University of Illinois at Urbana-Champaign, as a successor of METASAN, see [414]. ULTRASAN has been written in C++ (user interface) and C, making use of the X11 environment, and runs on SUN, DEC, RS/6000 and CONVEX systems.

Model class and evaluation techniques ULTRASAN, like METASAN, is based on stochastic activity networks. Models are input graphically, in a hierarchical fashion. Submodels can be defined which are subsequently used (one or more times) in higher-level models by use of the so-called replication-operator. Submodels of various kinds can be combined in a higher-level model by means of the join-operator. The joining and replicating of submodels can be done iteratively. Apart from exponentially timed activities, deterministically timed transitions are also allowed, albeit in a restricted fashion [291].

Steady-state measures are solved using either direct or iterative methods. In the former case a variant of the well-known LU-decomposition algorithm is used which reduces fill-in by a heuristic technique for pivot selection and which deals in a special way with very small elements. In the latter case a method based on successive over-relaxation is used. For transient and cumulative measures the randomization method is used. It yields transient state probabilities, expected values of cumulative sums as well as probability distributions. For the distribution of the accumulated reward during a fixed time interval, a randomization procedure has been developed that utilizes so-called path truncation to keep memory and cpu requirements small [378]. Also, for all the above measures, rate and impulse rewards can be combined.

All the analytical and numerical techniques use the reduced base model for their analysis, i.e., by making use of the structure of the model, in terms of submodels, joins and replications, state-space reductions of several orders of magnitude can be achieved [413].

ULTRASAN also provides steady-state and transient discrete-event simulations. By making use of the structure of the SAN model to be analyzed, very efficient simulation techniques have been implemented, in which a single event-list is exchanged for an event-tree [414]. This event-tree follows the tree-form of the model itself in terms of submodels that are iteratively joined and replicated. For large models, this structuring of the event-list increases the efficiency of the simulation. A technique based on importance sampling is used to speed up the simulations [366].

Model input and output   ULTRASAN makes use of a graphical interface. The SAN submodels are input with the SAN editor. The timings associated with the activities are input via menus. The activity rates and output gate functions are described in C. Once the submodels are defined they can be composed into an overall model with the graphical composed-model editor. Subeditors for the join and replicate functions are provided. The performability measures of interest are specified using a measure editor.

Once a model has been totally specified ULTRASAN generates a solution program which upon execution yields the desired measures in tabular form. The generated program can be a simulation program or a program for any of the included numerical techniques. A multiple run option is also provided. This allows the modeller to define series of experiments by defining global variables and indicating their value-ranges. For every possible combination of parameter values, ULTRASAN generates the corresponding solution program. These solution programs are then executed on a cluster of workstations in parallel.

GMTF fitting   With ULTRASAN the modelling activities totally take place at the SAN level, i.e., we have $\mathcal{F}_1$-modelling. The output is given in a form that can be understood at the level of model description, i.e., in the $\mathcal{R}_1$-domain. Thus, $\mathcal{V}_1$-evaluations can certainly be done with the tool. Since ULTRASAN allows for the easy construction of hierarchical models, higher-level evaluations are also possible. The rewards are input at the SAN level and need to be expressed as a general C function over the markings and/or the activity completions. Consequently, the rewards are not derived from a performance model (see the comments made with METASAN). Thus, $\mathcal{V}_i$-evaluations ($i \geq 1$) are possible with ULTRASAN.

Evaluation ULTRASAN provides a very general framework for building performability models. The tool supports a wide variety of numerical and simulative evaluation techniques. Since performance aspects as well as dependability aspects are dealt with, the potential of the tool is very large. The multiple run option allows for an easy execution of parametric studies.

10.5 Summary and outlook

For evaluation of the combined performance and dependability of fault-tolerant and distributed computer and communication systems, software tools are needed that support both the model construction as well as the model solution by one of the many mathematical techniques that have been developed for that purpose.

We have introduced a general modelling tool framework in Section 10.2 that can guide the design of and the thinking about tools. From this framework, some general rules for the structuring of performability software tools have been derived. Then in Section 10.3, we briefly stated which are the classes of measures to be obtained from a performability model, and in Section 10.4 we evaluated 17 software tools that can be used for performability modelling and evaluation. From this survey we can draw the following conclusions:

1. Over the years there has been a shift from lower-level model formalisms, i.e., normally the Markov reward models, to various higher-level formalisms.

2. This shift is still more apparent in the dependability model part than in the performance model part (rewards) and the output part.

3. Only the tools NUMAS, PENPET, SHARPE and DYQNTOOL$^+$ automatically derive and insert the rewards from a *model*.

4. In the SPN-based tools reward rates can be specified "at the net level", i.e., via simple numerical expressions over the markings but not via a model.

5. Tools based on SPNs have proved to be very flexible and generally applicable for performability modelling.

6. The object-oriented approach to systems modelling, as supported by TANGRAM-II, seems to be very general and easily extendible, since it is open ended towards various modelling approaches.

7. Tools based on SPAs, especially when exploiting compositionality, seem to have a high potential as well.

As noted before, a problem with any tool survey is related to completeness. We are glad to see many activities in this field, and therefore new tools might have emerged or existing tools might have been improved by the time this survey is published.

There is still much work to do in the field of performability modelling and evaluation. We just emphasize here some important research areas that are especially related to tools for performability evaluation (without intending to be exhaustive):

1. Tool support for hierarchical performability modelling formalisms that allows users to tailor their tool towards specific application domains.

2. Tool support for the high-level specification of the rewards (rate- and/or impulse-based) to be used.

3. Tool support for the specification of complex measures, in particular by using stochastic logics to specify measures (see the recent work in [18, 19, 211]) and to specify path-based measures [367]; in all these cases, the type of measure is exploited in the state-space generation process, that is, given a high-level model, an underlying state-space is derived that is just detailed enough to evaluate the required measures.

4. Techniques to exploit the similarity in the series of performance models to be solved for obtaining the rewards.

5. Tool support for automatically handling very large models, e.g., by using (approximate) truncation heuristics [202, 114], exact lumping techniques [96], state-space aggregation techniques [339], folding techniques [238], fixed-point iteration techniques [84, 76] and structured analysis techniques based on tensor algebra [56, 262].

6. Tool support for non-exponential timing, such as has been performed for DSPNexpress, SPNP and TimeNET.

7. Tool support for optimization purposes, as proposed for Penelope.

8. Tool support for parallel and distributed evaluation of large performability models, as discussed in e.g., [10, 64, 79, 210, 266].

9. Increasing the awareness among practising engineers of the need for performability modelling and evaluation.

Acknowledgements

Many of the tool designers have helped us in writing this chapter by providing specific information not available in the open literature. Their help is hereby acknowledged. The anonymous reviewers of this chapter are thanked for their constructive comments.

Bibliography

[1] H. Abdallah and R. Marie, "The uniformized power method for transient solutions of Markov processes", *Computers & Operations Research* 20(5), pp. 515–526, 1993.

[2] I.J.B.F. Adan and J. van der Wal, *Monotonicity of the Throughput in Single Server Production and Assembly Networks with Respect to Buffer Sizes, Queueing Networks with Blocking*, North-Holland, pp. 345–356, 1989.

[3] I.J.B.F. Adan and J. van der Wal, "Monotonicity of the throughput of a closed queueing network in the number of jobs", *Operations Research* 37, pp. 935–957, 1989.

[4] M. Ajmone Marsan, G. Balbo, A. Bobbio, G. Chiola, G. Conte and A. Cumani, "The effect of execution policies on the semantics and analysis of stochastic Petri nets", *IEEE Transactions on Software Engineering* 15, pp. 832–846, 1989.

[5] M. Ajmone Marsan, G. Balbo and G. Conte, "A class of generalized stochastic Petri nets for the performance evaluation of multiprocessor systems", *ACM Transactions on Computer Systems* 2, pp. 93–122, 1984.

[6] M. Ajmone Marsan and G. Chiola, "On Petri nets with deterministic and exponentially distributed firing times", in *Lecture Notes in Computer Science* 266, Springer Verlag, pp. 132–145, 1987.

[7] C. Alexopoulos and B.C. Shultes, "The balanced likelihood ratio method for estimating performance of highly reliable systems", in

Proceedings of the 1998 Winter Simulation Conference, IEEE Computer Society Press, pp. 1479–1486, 1998.

[8] C. Alexopoulos and B.C. Shultes, "Estimating reliability measures for highly dependable Markovian systems using balanced likelihood ratios", Research Report, School of Industrial and Systems Engineering, Georgia Institute of Technology, Atlanta, Georgia, 1999.

[9] S. Allmaier and S. Dalibor, "PANDA—Petri Net Analysis and Design Assistant", in *Tools Descriptions of the 9th International Conference on Modelling Techniques and Tools for Computer Performance Evaluation*, pp. 58–60, Saint Malo, France, 1997.

[10] S. Allmaier, M. Kowarschik and G. Horton, "State-space construction and steady-state solution of GSPNs on a shared-memory multiprocessor", in *Proceedings of the 7th International Workshop on Petri Nets and Performance Models*, pp. 112–121, IEEE Computer Society Press, 1997.

[11] W.A. Al-Qaq, M. Devetsikiotis and J.K. Townsend, "Importance sampling methodologies for simulation of communication systems with adaptive equalizers and time varying channels", *IEEE Journal on Selected Areas in Communications* 11, pp. 317–327, 1993.

[12] H.H. Ammar, S.M.R. Islam and S. Deng, "Performability analysis of parallel and distributed algorithms", in *Proceedings of the International Workshop on Petri Nets and Performance Models*, IEEE Computer Society Press, pp. 240–248, 1989.

[13] S. Andradottir, D.P. Heyman and T.J. Ott, "On the choice of alternative measures in importance sampling with Markov chains", *Operations Research* 43(3), pp. 509–519, 1995.

[14] S. Asmussen and R.Y. Rubinstein, "Steady state rare event simulation in queueing models and its complexity properties", in *Advances in Queueing: Models, Methods and Problems*, J. Dshalalow (ed.), CRC Press, pp. 429–466, 1994.

[15] B.E. Aupperle and J.F. Meyer, "Fault-tolerant BIBD networks", in *Proceedings of the 18th International Symposium on Fault-Tolerant Computing*, pp. 306–311, Tokyo, Japan, June 1988.

[16] B.E. Aupperle, J.F. Meyer and L. Wei, "Evaluation of fault-tolerant systems with nonhomogeneous workloads", in *Proceedings of the 19th International Symposium on Fault-Tolerant Computing*, pp. 159–166, Chicago, IL, June 1989.

[17] B. Avi-Itzhak and P. Naor, "Some queueing problems with the service station subject to breakdowns", *Operations Research* 11, pp. 303–320, 1963.

[18] C. Baier, B.R. Haverkort, H. Hermanns and J.-P. Katoen, "Model checking continuous-time Markov chains by transient analysis", in *Computer-Aided Verification, Lecture Notes in Computer Science* 1855, Springer Verlag, pp. 358–372, 2000.

[19] C. Baier, B.R. Haverkort, H. Hermanns and J.-P. Katoen, "On the logical characterisation of performability properties", in *Automata, Languages, and Programming, Lecture Notes in Computer Science* 1853, Springer Verlag, pp. 780–792, 2000.

[20] M.O. Ball, C.J. Colbourn and J.S. Provan, "Network reliability", in *Handbook of Operations Research: Network Models*, North-Holland, pp. 673–762, 1995.

[21] R.E. Barlow and F. Proschan, *Statistical Theory of Reliability and Life Testing*, Holt, Rinehart and Winston, Maryland, 1981.

[22] R.E. Barlow and F. Proschan, *Statistical Theory of Reliability and Life Testing*, Holt, Rinehart and Winston, New York, 1975.

[23] F. Bause, P. Buchholz and P. Kemper, "QPN-Tool for the specification and analysis of hierarchically combined queueing Petri nets", in *Quantitative Evaluation of Computing and Communication Systems*, H. Beilner, F. Bause, editors, *Lecture Notes in Computer Science* 977, Springer Verlag, pp. 224–238, 1995.

[24] F. Bause, P. Buchholz and P. Kemper, "Integrating software and hardware performance models using hierarchical queueing Petri nets", in *Proceedings of the 9th ITG/GI-Fachtagung Messung, Modellierung und Bewertung von Rechen- und Kommunikationssystemen*, pp. 97–101, VDE Verlag, 1997.

[25] F. Bause and P.S. Kritzinger, *Stochastic Petri Nets: An Introduction to the Theory*, Vieweg Verlag, 1996.

[26] A.J. Bayes, "Statistical techniques for simulation models", *Australian Computer Journal* 2, pp. 180–184, 1970.

[27] M.D. Beaudry, "Performance related reliability measures for computing systems", *IEEE Transactions on Computers* 27(6), pp. 540–547, 1978.

[28] J. Bechta Dugan, K.S. Trivedi, R. Geist and V.F. Nicola, "Extended stochastic Petri nets: application and analysis", in *Proceedings Performance'84*, North-Holland, 1985.

[29] H. Beilner, "Workload characterization and performance modelling tools", *Proceedings of the International Workshop on Workload Characterization of Computer Systems*, Pavia, Italy, October 23–25, 1985.

[30] H. Beilner, J. Mäter and N. Weissenberg, "Towards a performance modelling environment: news on HIT", in *Modelling Techniques and Tools for Computer Performance Evaluation*, D. Potier, R. Puigjaner, editors, Plenum Press, pp. 57–75, 1989.

[31] C. Béounes, M. Aguéra, J. Arlat, S. Bachmann, C. Bourdeau, J.-E. Doucet, K. Kanoun, J.-C. Laprie, S. Metge, J. Moreira de Souza, D. Powell and P. Spiesser, "SURF-2: a program for dependability evaluation of complex hardware and software systems", *Proceedings FTCS* 23, IEEE Computer Society Press, pp. 668–673, 1993.

[32] M. Bernardo, *Theory and Application of Extended Markovian Process Algebra*, Ph.D. thesis, University of Bologna, 1999.

[33] M. Bernardo and R. Gorrieri, "A tutorial on EMPA: a theory of concurrent processes with nondeterminism, priorities, probabilities and time", *Theoretical Computer Science* 202(1&2), pp. 1–54, 1998.

[34] S. Berson, E. de Souza e Silva and R.R. Muntz, "An Object Oriented Methodology for the Specification of Markov Models", *UCLA Technical Report* CSD-870030, 1987.

[35] D.P. Bertsekas, *Dynamic Programming: Deterministic and Stochastic Models*, Prentice Hall, Englewood Cliffs, NJ, 1987.

[36] A. Blum, A. Goyal, P. Heidelberger, S.S. Lavenberg, M.K. Nakayama and P. Shahabuddin, "Modeling and analysis of system dependability using the system availability estimator", *Proceedings of the 24th International Symposium on Fault Tolerant Computing*, IEEE Computer Society Press, pp. 137–141, 1994.

[37] A. Bobbio and K.S. Trivedi, "An aggregation technique for the transient analysis of stiff Markov chains", *IEEE Transactions on Computers* 35(9), pp. 803–814, 1986.

[38] A. Bobbio, V.G. Kulkarni, A. Puliafito, M. Telek and K.S. Trivedi, "Preemptive repeat identical transitions in Markov Regenerative Stochastic Petri Nets", in *Proceedings 6th International Conference on Petri Nets and Performance Models*, IEEE Computer Society Press, pp. 113–122, 1995.

[39] A. Bobbio, "The effect of an imperfect coverage on the optimum degree of redundancy of a degradable multiprocessor system", in *Proceedings Reliability'87*, Paper 5B/3, Birmingham, 1987.

[40] A. Bobbio, "Petri nets generating Markov reward models for performance/reliability analysis of degradable systems", in *Modeling Techniques and Tools for Computer Performance Evaluation*, R. Puigjaner and D. Potier, editors, Plenum Press, pp. 353–365, 1989.

[41] A. Bobbio, "A multi-reward stochastic model for the completion time of parallel tasks", *in Teletraffic and Datatraffic, Proceedings 13th International Teletraffic Congress*, A. Jensen and V.B. Iversen, editors, North-Holland, pp. 577–582, 1991.

[42] A. Bobbio, A. Puliafito and M. Telek. "A modeling framework to implement preemption policies in non-Markovian SPN", *IEEE Transactions on Software Engineering* 26, pp. 353–365, 2000.

[43] A. Bobbio, A. Puliafito, M. Telek and K. Trivedi, "Recent developments in non-Markovian stochastic Petri nets", *Journal of Systems Circuits and Computers*, 8(1) pp. 119–158, 1998.

[44] A. Bobbio and L. Roberti, "Distribution of the minimal completion time of parallel tasks in multi-reward semi-Markov models", *Performance Evaluation* 14, pp. 239–256, 1992.

[45] A. Bobbio and M. Telek, "Computational restrictions for SPN with generally distributed transition times", in *Proceedings of First European Dependable Computing Conference*, D. Hammer, K. Echtle and D. Powell, editors, *Lecture Notes in Computer Science* 852, Springer Verlag, pp. 131–148, 1994.

[46] A. Bobbio and M. Telek, "Markov regenerative SPN with non-overlapping activity cycles", in *International Computer Performance and Dependability Symposium*, IEEE CS Press, pp. 124–133, 1995.

[47] A. Bobbio and M. Telek. "Non-exponential stochastic Petri nets: an overview of methods and techniques", *Computer Systems: Science & Engineering* 13(6), pp. 339–351, 1998.

[48] A. Bobbio and K.S. Trivedi, "Computation of the distribution of the completion time when the work requirement is a PH random variable", *Stochastic Models* 6, pp. 133–149, 1990.

[49] M. Bouissou, "The FIGARO dependability evaluation workbench in use: case studies for fault-tolerant computer systems", in *Proceedings FTCS* 23, IEEE Computer Society Press, pp. 680–685, 1993.

[50] W.G. Bouricius, W.C. Carter and P.R. Schneider, "Reliability modeling techniques for self-repairing computer systems", in *Proceedings of ACM National Conference*, San Francisco, CA, pp. 295–309, 1969.

[51] P. Bratley, B.L. Fox and L.E. Schrage, *A Guide to Simulation*, Second Edition, Springer Verlag, New York, 1987.

[52] L. Breiman, *Probability*, Addison-Wesley, Reading, MA, 1968.

[53] C.E.F. de Brito, E. de Souza e Silva, M.C. Diniz and R.M.M. Le ao, "Transient analysis of multimedia source models", in *Proceedings of the 18th Brazilian Computer Networks Symposium*, pp. 519–534, May 2000.

[54] M. Brown, "Error bounds for exponential approximations of geometric convolutions", *The Annals of Probability* 18(3), pp. 1388–1402, 1990.

[55] P. Buchholz, "Exact and ordinary lumpability in finite Markov chains", *Journal of Applied Probability* 31, pp. 59–75, 1994.

[56] P. Buchholz, "Hierarchical structuring of superposed GSPNs", in *Proceedings of the 7th International Workshop on Petri Nets and Performance Models*, IEEE Computer Society Press, pp. 81–90, 1997.

[57] R.M.L.R. Carmo, L.R. de Carvalho, E. de Souza e Silva, M.C. Diniz and R.R. Muntz, "TANGRAM II: a performability modeling environment tool", in *Computer Performance Evaluation, Modelling Techniques and Tools*, R. Marie, B. Plateau, M. Calzarossa, G. Rubino, editors, *Lecture Notes in Computer Science* 1245, Springer-Verlag, pp. 6–18, 1997.

[58] R.M.L.R. Carmo, L.R. de Carvalho, E. de Souza e Silva and R.R. Muntz, "Performance/availability modeling with the TANGRAM II modeling environment", *Performance Evaluation* 33, pp. 45–65, 1998.

[59] J.A. Carrasco and A. Calderon, "Regenerative randomization: theory and application examples", *Performance Evaluation Review* 23(1), pp. 241–252, May 1995.

[60] J.A. Carrasco and J. Figueras, "METFAC: design and implementation of a software tool for modeling and evaluation of complex fault-tolerant computing systems", *Proceedings of the 16th International Fault-Tolerant Computing Symposium*, IEEE Computer Society Press, pp. 424–429, 1986.

[61] J.A. Carrasco, *Modelación y Evaluación de la Tolerancia a Fallos de Sistemas Distribuidos con Capacidad de Reconfiguracion*, Ph.D. thesis, University of Catalunya, Spain, 1986.

[62] J.A. Carrasco, "Failure distance-based simulation of repairable fault-tolerant systems", in *Proceedings of the Fifth International Conference on Modeling Techniques and Tools for Computer Performance Evaluation*, North Holland, Amsterdam, pp. 337–351, 1992.

[63] J.A. Carrasco, "Efficient transient simulation of failure/repair Markovian models", in *Proceedings of the Tenth Symposium on Reliable and Distributed Computing*, IEEE Computer Society Press, pp. 152–161, 1991.

[64] S. Caselli, G. Conte and P. Marenzoni, "Parallel state space exploration for GSPN models" in *Applications and Theory of Petri Nets*, G. De Michelis and M. Diaz, editors, *Lecture Notes in Computer Science* 935, Springer-Verlag, pp. 181–200, 1995.

[65] X. Castillo and D.P. Siewiorek, "A performance reliability model for computing systems", in *Proceedings 10th International Symposium on Fault-Tolerant Computing*, pp. 187–192, 1980.

[66] *CCITT Blue Book, Fascicle III.1*, International Telecommunication Union, Geneva, 1989.

[67] C.S. Chang, "Stability, queue length and delay of deterministic and stochastic queueing networks", *IEEE Transactions on Automatic Control* 39, pp. 913–931, 1994.

[68] C.S. Chang, P. Heidelberger, S. Juneja and P. Shahabuddin, "Effective bandwidth and fast simulation of ATM in-tree networks", *Performance Evaluation* 20, pp. 45–65, 1994. Also in *Proceedings of the Performance '93 Conference*.

[69] S. C. Chapra and R. P. Canale, *Numerical Methods for Engineers*, McGraw-Hill, New York, 1988.

[70] P.F. Chimento and K.S. Trivedi, "The completion time of programs on processors subject to failure and repair", *IEEE Transactions on Computers* 42, pp. 1184–1194, 1993.

[71] G. Chiola, "A graphical Petri net tool for performance analysis", in *Modelling Techniques and Performance Evaluation*, S. Fdida and G. Pujolle, editors, North-Holland, pp. 323–333, 1987.

[72] G. Chiola, "Simulation framework for timed and stochastic Petri nets", *International Journal in Computer Simulation* 1, pp. 153–168, 1991.

[73] G. Chiola, C. Dutheillet, G. Franceschines and S. Haddad, "Stochastic well-formed colored nets and symmetric modelling applications", *IEEE Transactions on Software Engineering* 42(11), pp. 1343–1360, 1993.

[74] H. Choi, V.G. Kulkarni and K. Trivedi, "Transient analysis of deterministic and stochastic Petri nets", in *Proceedings of the 14th International Conference on Application and Theory of Petri Nets*, Chicago, June 1993.

[75] H. Choi, V.G. Kulkarni and K. Trivedi, "Markov regenerative stochastic Petri nets", *Performance Evaluation* 20, pp. 337–357, 1994.

[76] H. Choi and K.S. Trivedi, "Approximate performance models of polling systems using stochastic Petri nets", *Proceedings INFO-COM'92*, IEEE Computer Society Press, pp. 2306–2314, 1992.

[77] G. Ciardo, A. Blakemore, P.F.J. Chimento, J.K. Muppala and K.S. Trivedi, "Automated generation and analysis of Markov reward models using stochastic reward nets", in *Linear Algebra, Markov Chains and Queueing Models*, C. Meyer and R.J. Plemmons, editors, *IMA Volumes in Mathematics and Its Applications* 48, Springer-Verlag, pp. 145–191, 1993.

[78] G. Ciardo, R. German and C. Lindemann, "A characterization of the stochastic process underlying a stochastic Petri net", *IEEE Transactions on Software Engineering* 20, pp. 506–515, 1994.

[79] G. Ciardo, J. Gluckman and D. Nicol, "Distributed state space generation of discrete-state stochastic models", *INFORMS Journal on Computing* 10(1), pp. 82–93, 1998.

[80] G. Ciardo and C. Lindemann, "Analysis of deterministic and stochastic Petri nets", in *Proceedings International Workshop on Petri Nets*

and Performance Models, IEEE Computer Society Press, pp. 160–169, 1993.

[81] G. Ciardo, R. Marie, B. Sericola and K.S. Trivedi, "Performability analysis using semi-Markov reward processes", *IEEE Transactions on Computers* 39(10), pp. 1251–1264, 1990.

[82] G. Ciardo, J. Muppala and K.S. Trivedi, "SPNP: Stochastic Petri Net Package", in *Proceedings of the International Workshop on Petri Nets and Performance Models*, IEEE Computer Society Press, pp. 142–151, 1989.

[83] G. Ciardo, J.K. Muppala and K.S. Trivedi, "On the solution of GSPN reward models", *Performance Evaluation* 12(4), pp. 237–254, 1991.

[84] G. Ciardo and K.S. Trivedi, "A decomposition approach for stochastic reward net models", *Performance Evaluation* 18, pp. 37–59, 1993.

[85] G. Ciardo and R. Zijal, "Well-defined stochastic Petri nets", in *Proceedings MASCOTS 1996*, San Jose, CA, 1996.

[86] B. Ciciani and V. Grassi, "Performability evaluation of fault-tolerant satellite systems", *IEEE Transactions on Communications* 35(4), pp. 403–409, 1987.

[87] E. Çinlar, *Introduction to Stochastic Processes*, Prentice-Hall, Englewood Cliffs, New Jersey, 1975.

[88] E. Çinlar, "Markov renewal theory", *Advances in Applied Probability* 1, pp. 123–187, 1969.

[89] G. Clark, S. Gilmore, J. Hillston and M. Ribaudo, "Exploiting model logic to express performance measures", in *Computer Performance Evaluation: Modelling Techniques and Tools*, B.R. Haverkort, H. Bohnenkamp, C.U. Smith, editors, *Lecture Notes in Computer Science* 1786, Springer Verlag, pp. 247–261, 2000.

[90] R. Cogburn, "A uniform theory for sums of Markov chain transition probabilities", *The Annals of Probability* 3, pp. 191–214, 1975.

[91] A.E. Conway and A. Goyal, "Monte Carlo simulation of computer system availability/reliability models", in *Proceedings of the Seventeenth Symposium on Fault-Tolerant Computing*, IEEE Computer Society Press, pp. 230–235, 1987.

[92] A. Costes, J.-E. Doucet, C. Landrault and J.-C. Laprie, "SURF: a program for dependability evaluation of complex fault-tolerant computing systems", in *Proceedings of the 11th International Fault-Tolerant Computing Symposium*, IEEE Computer Society Press, pp. 72–78, 1981.

[93] M. Cottrell, J.C. Fort and G. Malgouyres, "Large deviations and rare events in the study of stochastic algorithms", *IEEE Transactions on Automatic Control* 28, pp. 907–920, 1983.

[94] P.-J. Courtois, *Decomposability, Queueing and Computer Science Applications*, Academic Press, 1977.

[95] P.-J. Courtois and P. Semal, "Computable bounds for conditional steady-state probabilities in large Markov chains and queueing models", *IEEE Journal on Selected Areas in Communications* 46, Sept 1996.

[96] J. Couvillion, R. Freire, R. Johnson, W.D. Obal II, M.A. Qureshi, M. Rai, W.H. Sanders and J. Tvedt, "Performability modeling with *UltraSAN*", *IEEE Software* 8(5), pp. 69–80, 1991.

[97] D.R. Cox, "The analysis of non-markovian stochastic processes by the inclusion of supplementary variables", in *Proceedings of the Cambridge Philosophical Society* 51, pp. 433–440, 1955.

[98] M.A. Crane and D.L. Iglehart, "Simulating stable stochastic systems, III: Regenerative processes and discrete-event simulations", *Operations Research* 23, pp. 33–45, 1975.

[99] A. Csenki, "A dependability measure for Markov models of repairable systems: Solution by randomization and computational experience", *Computers and Mathematics with Applications* 30(2), pp. 95–110, 1995.

[100] A. Cumani, "Esp – a package for the evaluation of stochastic Petri nets with phase-type distributed transition times", in *Proceedings International Workshop Timed Petri Nets*, IEEE Computer Society Press, 674, pp. 144–151, Torino, Italy, 1985.

[101] O. Das and C.M. Woodside, "The fault-tolerant layered queueing network model for performability of distributed systems", in *Proceedings of the 3rd IEEE International Computer Performance and Dependability Symposium*, IEEE CS Press, pp. 132–141, 1998.

[102] M.H.A. Davis, "Piecewise deterministic Markov processes: A general class of non-diffusion stochastic models", *Journal of the Royal Statistical Society B.* 46, pp. 353–388, 1984.

[103] E. de Souza e Silva and H.R. Gail, "Calculating cumulative operational time distributions of repairable computer systems", *IEEE Transactions on Computers* 35(4), pp. 322–332, 1986.

[104] E. de Souza e Silva and H.R. Gail, "Calculating availability and performability measures of repairable computer systems using randomization", *Journal of the ACM* 36(1), pp. 171–193, 1989.

[105] E. de Souza e Silva and H.R. Gail, "Analyzing scheduled maintenance policies for repairable computer systems", *IEEE Transactions on Computers* 39(11), pp. 1309–1324, 1990.

[106] E. de Souza e Silva and H.R. Gail, "Performability analysis of computer systems: from model specification to solution", *Performance Evaluation* 14(3&4), pp. 157–196, 1992.

[107] E. de Souza e Silva and H.R. Gail, "An algorithm to calculate transient distributions of cumulative reward", *UCLA Technical Report CSD-940021*, May 1994.

[108] E. de Souza e Silva, H.R. Gail and R. Vallejos Campos, "Calculating transient distribution of cumulative reward", *Performance Evaluation Review* 23(1), pp. 231–240, May 1995.

[109] E. de Souza e Silva and H.R. Gail, "An algorithm to calculate transient distributions of cumulative rate and impulse based reward",

Communications in Statistics – Stochastic Models 14(3), pp. 509–536, 1998.

[110] E. de Souza e Silva and H.R. Gail, "Transient solutions for Markov chains", in *Computational Probability*, W. Grassmann, editor, Kluwer Academic Publishers, pp. 43–81, 2000.

[111] E. de Souza e Silva, H.R. Gail and J.C. Guedes, "Transient distributions of cumulative rate and impulse based reward with applications", in *System Modelling and Optimization*, M.P. Polis *et al.* , editors, CRC Press, pp. 298–306, 1999.

[112] E. de Souza e Silva, H.R. Gail and R.R. Muntz, "Efficient solutions for a class of non-Markovian models", in *Computations with Markov Chains*, W.J. Stewart, editor, Kluwer Academic Publishers, pp. 483–506, 1995.

[113] E. de Souza e Silva, R.M.M Le ao and M.C. Diniz, "Transient analysis applied to traffic modeling", in *2nd Workshop on Mathematical Modeling and Analysis*, 2000.

[114] E. de Souza e Silva and P.M. Ochoa, "State space exploration in Markov models", *ACM Performance Evaluation Review* 20(1), pp. 152–166, 1992.

[115] G. de Veciana, C. Courcoubetis and J. Walrand, "Decoupling bandwidth for networks: a decomposition approach for resource management for networks", In *IEEE INFOCOM'94 Proceedings*, IEEE Computer Society Press, pp. 466–473, 1993.

[116] M. Devetsikiotis and J.K. Townsend, "Statistical optimization of dynamic importance sampling parameters for efficient simulation of communication networks", *IEEE/ACM Transactions on Networking* 1, pp. 293–305, 1993.

[117] J. Dieudonné, *Foundation of Modern Analysis, vol. I*, Academic Press, New York, 1969.

[118] N.M. van Dijk, "Simple bounds for queueing systems with breakdowns", *Performance Evaluation* 8, pp. 117–128, 1988.

[119] N.M. van Dijk, "Simple throughput bounds for large queueing networks with finite capacity constraints", *Peformance Evaluation* 9, pp. 153–167, 1989.

[120] N.M. van Dijk, "On a simple proof of uniformization for continuous and discrete-state continuous-time Markov chains", *Advances in Applied Probability* 22, pp. 749–750, 1990.

[121] N.M. van Dijk, "On the importance of bias-terms for error bounds and comparison results", in *Numerical Solutions of Markov Chains*, W.J. Stewart, editor, Marcel Dekker, New York, pp. 617–642, 1991.

[122] N.M. van Dijk, "Transient error bounds analysis for continuous-time Markov reward structures", *Performance Evaluation* 13, pp. 147–158, 1991.

[123] N.M. van Dijk, "Approximate uniformization for continuous-time Markov chains with an application to performability analysis", *Stochastic Processes and Their Applications* 40(2), pp. 339–357, 1992.

[124] N.M. van Dijk, "Uniformization for nonhomogeneous Markov chains", *Operations Research Letters* 12, pp. 283–291, 1992.

[125] N.M. van Dijk, B.R. Haverkort and I.G. Niemegeers, "Performability modelling of computer and communication systems", *Performance Evaluation* 14(3&4), pp. 135–138, 1992.

[126] N.M. van Dijk and M. Miyazawa, "Error bounds on a practical approximation for finite tandem queues", *Operations Research Letters*, 21, pp. 201–208, 1997.

[127] N.M. van Dijk and M. Miyazawa, "A note on bounds and error bounds for nonexponential batch arrival systems", *Probability in the Engineering and Informational Sciences* 11, pp. 189–201, 1997.

[128] N.M. van Dijk and P.G. Taylor, "Strong stochastic bounds for the stationary distribution of a class of multicomponent performability models", *Operations Research* 46, pp. 665–674, 1998.

[129] N.M. van Dijk, P. Tsoucas and J. Walrand, "Simple bounds and monotonicity of the call congestion of infinity multiserver delay systems", *Probability in the Engineering and Informational Sciences* 2, pp. 129–138, 1988.

[130] N.M. van Dijk and J. van der Wal, "Simple bounds and monotonicity results for multi-server exponential tandem queues", *Queueing Systems* 4, pp. 1–16, 1989.

[131] N.M. van Dijk and H. Korezlioglu, "On product form approximation for communication networks with losses: error bounds", *Annals of Operations Research* 35, pp. 60–94, 1992.

[132] N.M. van Dijk and M.L. Puterman, "Perturbation theory for Markov reward processes with applications to queueing systems", *Advances in Applied Probability* 20, pp. 79–89, 1988.

[133] J.D. Diener and W.H. Sanders, "Empirical comparison of uniformization methods for continuous-time Markov chains", in *Computations with Markov Chains*, W.J. Stewart, editor, Kluwer Academic Publishers, pp. 547–570, 1995.

[134] M.C. Diniz, *Solution Techniques for Models Arising from Multimedia Networks* (in Portuguese), Ph.D. Thesis, COPPE/Sistemas, Federal University of Rio de Janeiro, 2000.

[135] L. Donatiello and V. Grassi, "On evaluating the cumulative performance distribution of fault-tolerant computer systems", *IEEE Transactions on Computers* 40(11), pp. 1301–1307, 1991.

[136] L. Donatiello and B.R. Iyer, "Analysis of a composite performance reliability measure for fault-tolerant systems", *Journal of the ACM* 34(1), pp. 179–199, 1987.

[137] J. B. Dugan, *Extended Stochastic Petri Nets: Applications and Analysis*, Ph.D. Thesis, Duke University, 1984.

[138] J.B. Dugan, K.S. Trivedi, M.K. Smotherman and R.M. Geist, "The hybrid automated reliability predictor", *Journal of Guidance, Control and Dynamics* 9(3), pp. 319–331, 1986.

[139] A. Von Ellenrieder and A. Levine, "The probability of an excessive non-functioning interval", *Operations Research* 14, pp. 835–840, 1966.

[140] M. Falkner, M. Devetsikiotis and I. Lambadaris, "Fast simulation of networks of queues using effective and decoupling bandwidths", *ACM Transactions on Modeling and Computer Simulation* 9, pp. 45–58, 1999.

[141] D. Ferrari, *Computer Systems Performance Evaluation*, Prentice-Hall, 1978.

[142] G.S. Fishman, "A Monte-Carlo sampling plan for evaluating network reliability", *Operations Research* 34(4), pp. 581–594, 1986.

[143] B.L. Fox and P.W. Glynn, "Discrete-time conversion for simulating semi-Markov processes", *Operations Research Letters* 5, pp. 191–196, 1986.

[144] B.L. Fox and P.W. Glynn, "Computing Poisson probabilities", *Communications of the ACM* 31(4), pp. 440–445, 1988.

[145] B.L. Fox and P.W. Glynn, "Discrete-time conversion for simulating finite-horizon Markov processes", *SIAM Journal of Applied Mathematics* 50(5), pp. 1457–1473, 1990.

[146] B.L. Fox and P.W. Glynn, "Estimating time averages via randomly-spaced observations", *Probability in the Engineering and Informational Sciences* 3, pp. 299–318, 1989.

[147] B.L. Fox and P.W. Glynn, "Replication schemes for limiting expectations", *SIAM Journal of Applied Mathematics* 47, pp. 186–214, 1989.

[148] M.R. Frater, T.M. Lennon and B.D.O. Anderson, "Optimally efficient estimation of the statistics of rare events in queueing networks", *IEEE Transactions on Automatic Control* 36, pp. 1395–1405, 1991.

[149] K.O. Friedrichs, "Symmetric hyperbolic linear differential equations", *Communications of Pure and Applied Mathematics* 7, pp. 345–392, 1954.

[150] M. Fukushima and M. Hitsuda, "On a class of Markov processes taking values on lines and the central limit theorem", *Nagoya Journal of Mathematics* 30, pp. 7–56, 1967.

[151] D.G. Furchtgott and J.F. Meyer, "A performability solution method for degradable non-repairable systems", *IEEE Transactions on Computers* 33(6), pp. 550–554, 1984.

[152] D.G. Furchtgott, *Performability Models and Solutions*, Ph.D. thesis, University of Michigan, 1984.

[153] D.P. Gaver, "A waiting line with interrupted service, including priorities", *Journal of the Royal Statistical Society B.* 24, pp. 73–90, 1962.

[154] F.A. Gay and M.L. Ketelsen, "Performance evaluation for gracefully degrading systems", in *Proceedings 9th International Fault-Tolerant Computing Symposium*, pp. 51–58, 1979.

[155] R.M. Geist and K.S. Trivedi, "Ultra-high reliability prediction for fault-tolerant computer systems", *IEEE Transactions on Computers* 32, pp. 1118–1127, 1983.

[156] R.M. Geist and M.K. Smotherman, "Ultrahigh reliability estimates through simulation", *Proceedings of the Annual Reliability and Maintainability Symposium*, IEEE Computer Society Press, pp. 350–355, 1989.

[157] R. Geist and K.S. Trivedi, "Reliability estimation of fault-tolerant systems: tools and techniques", *IEEE Computer Society Press* 23(6), pp. 52–61, 1990.

[158] R. German, *Analysis of Stochastic Petri Nets with Non-Exponentially Distributed Firing Times*, Ph.D. thesis, TU Berlin, 1994.

[159] R. German, "New results for the analysis of deterministic and stochastic Petri nets", in *International Computer Performance and Dependability Symposium (IPDS'95)*, IEEE Computer Society Press, pp. 114–123, 1995.

[160] R. German, "SPNL: processes as language oriented building blocks of stochastic Petri nets", in *Computer Performance Evaluation, Modelling Techniques and Tools*, R. Marie, B. Plateau, M. Calzarossa, G. Rubino, editors, *Lecture Notes in Computer Science* 1245, Springer Verlag, pp. 123–134, 1997.

[161] R. German, "Iterative analysis of Markov regenerative models", in *Computer Performance Evaluation: Modelling Techniques and Tools*, B. Haverkort, H. Bohnenkamp and C. Smith, editors, *Lecture Notes in Computer Science* 1786, Springer Verlag, pp. 156–170, 2000.

[162] R. German, C. Kelling, A. Zimmermann and G. Hommel, *TimeNET – A toolkit for evaluating non-Markovian stochastic Petri nets*, Technische Universität Berlin, 1994.

[163] R. German, Ch. Kelling, A. Zimmermann and G. Hommel, "TimeNET: a toolkit for evaluating non-Markovian Petri nets", *Performance Evaluation* 24(1&2), pp. 69–88, 1995.

[164] R. German and C. Lindemann, "Analysis of stochastic Petri nets by the method of supplementary variables", *Performance Evaluation* 20, pp. 317–336, 1994.

[165] R. German and J. Mitzlaff, "Transient analysis of deterministic and stochastic Petri nets with TimeNET", in *Quantitative Evaluation of Computing and Communication Systems*, H. Beilner, F. Bause, editors, *Lecture Notes in Computer Science* 977, Springer Verlag, pp. 209–223, 1995.

[166] S.B. Gershwin, "Variance of output of a tandem production system", Technical Report, Massachusetts Institute of Technology, Laboratory for Information and Decision Systems, January 1992.

[167] I.B. Gertsbakh, "Asymptotic methods in reliability theory: a review", *Advances in Applied Probability* 16, pp. 147–175, 1984.

[168] S. Gilmore and J. Hillston, "The PEPA workbench: a tool to support a process algebra-based approach to performance modelling", in *Lecture Notes in Computer Science* 794, Springer Verlag, pp. 353–368, 1994.

[169] P. Glasserman, P. Heidelberger, P. Shahabuddin and T. Zajic, "Splitting for rare event simulation: analysis of simple cases", *Proceedings of the 1996 Winter Simulation Conference*, IEEE Computer Society Press, pp. 302–308, 1996.

[170] P. Glasserman, P. Heidelberger, P. Shahabuddin and T. Zajic, "A large deviations perspective on the efficiency of multilevel splitting", *IEEE Transactions on Automatic Control* 43(12), pp. 1666–1679, 1998.

[171] P. Glasserman, P. Heidelberger, P. Shahabuddin and T. Zajic, "Multilevel splitting for estimating rare event probabilities", *Operations Research* 47(4), pp. 585–600, 1999.

[172] P. Glasserman and S.G. Kou, "Analysis of an importance sampling estimator for tandem queues", *ACM Transactions on Modeling and Computer Simulation* 5(1), pp. 22–42, 1995.

[173] P.W. Glynn, "A GSMP formalism for discrete event systems", in *Proceedings of the IEEE* 77, pp. 14–23, 1989.

[174] P.W. Glynn, "Likelihood ratio derivative estimators for stochastic systems", in *Proceedings of the 1989 Winter Simulation Conference*, IEEE Computer Society Press, pp. 374–380, 1989.

[175] P.W. Glynn, "Importance sampling for Markov chains: Asymptotics for the variance", *Stochastic Models* 10, pp. 701–717, 1995.

[176] P.W. Glynn, P. Heidelberger, V.F. Nicola and P. Shahabuddin, "Efficient estimation of steady-state measures in non-regenerative dependability models", *Proceedings of the 1993 Winter Simulation Conference*, IEEE Computer Society Press, pp. 311–316, 1993.

[177] P.W. Glynn and D.L. Iglehart, "Importance sampling for stochastic simulations", *Management Science* 35, pp. 1367–1392, 1989.

[178] G.H. Golub and C.F. Van Loan, *Matrix Computations*, 2nd edition, Johns Hopkins University Press, Baltimore, MD, 1989.

[179] V. Gopalakrishna, *Composite Performance-Reliability Analysis of Flexible Manufacturing Systems*, Ph.D. Thesis, Indian Institute of Science, Bangalore, June 1994.

[180] A. Goyal, W.C. Carter, E. de Souza e Silva, S.S. Lavenberg and K.S. Trivedi, "The system availability estimator", in *Proceedings of the 16th International Fault-Tolerant Computing Symposium*, IEEE Computer Society Press, pp. 84–89, 1986.

[181] A. Goyal, S.S. Lavenberg and K.S. Trivedi, "Probabilistic modelling of computer system availability", *Annals of Operations Research* 8, pp. 285–306, 1987.

[182] A. Goyal, "System Availability Estimator – User's Manual", Version 2.0, Internal report, IBM Yorktown Heights, 1987.

[183] A. Goyal and S.S. Lavenberg, "Modelling and analysis of computer system availability", *IBM Journal of Research and Development* 31(6), pp. 651–664, 1987.

[184] A. Goyal, P. Heidelberger and P. Shahabuddin, "Measure specific dynamic importance sampling for availability simulations", in *Proceedings of the 1987 Winter Simulation Conference*, IEEE Computer Society Press, pp. 351–357, 1987.

[185] A. Goyal, P. Shahabuddin, P. Heidelberger, V.F. Nicola and P.W. Glynn, "A unified framework for simulating Markovian models of highly dependable systems", *IEEE Transactions on Computers* 41(1), pp. 36–51, 1992.

[186] A. Goyal and A.N. Tantawi, "Evaluation of performability for degradable computer systems", *IEEE Transactions on Computers* 36(6), pp. 738–744, 1987.

[187] A. Goyal, and A.N. Tantawi, "A measure of guaranteed availability and its numerical evaluation", *IEEE Transactions on Computers* 37(1), pp. 25–32, 1988.

[188] V. Grassi, L. Donatiello and G. Iazeolla, "Performability evaluation of multicomponent fault-tolerant systems", *IEEE Transactions on Reliability* 37(2) pp. 216–222, 1988.

[189] W.K. Grassmann, "Transient solutions in Markovian queueing systems", *Computers & Operations Research* 4, pp. 47–53, 1977.

[190] W.K. Grassmann, "Transient solutions in Markovian queues", *European Journal of Operational Research* 1, pp. 396–402, 1977.

[191] W.K. Grassmann, "The GI/PH/1 queue: a method to find the transition matrix", *INFOR* 20(2), pp. 144–156, 1982.

[192] W.K. Grassmann, "Means and variances of time averages in Markovian environments", *European Journal of Operational Research* 31(1), pp. 132–139, 1987.

[193] W.K. Grassmann, "Numerical solutions for Markovian event systems", in *Quantitative Methoden in den Wirtschaftswissenschaften*, P. Kall *et al.*, editors, pp. 73–87, Springer Verlag, 1989.

[194] W.K. Grassmann, "Finding transient solutions in Markovian event systems through randomization", in *Numerical Solution of Markov Chains*, W. J. Stewart, editor, Marcel Dekker, New York, pp. 357–371, 1991.

[195] W.K. Grassmann, M.I. Taksar and D.P. Heyman, "Regenerative analysis and steady-state distributions for Markov chains", *Operations Research* 33(5), pp. 1107–1116, 1985.

[196] R.J. Griego and R. Hersh, "Random evolutions, Markov chains and systems of partial differential equations", in *Proceedings National Academy of Sciences USA* 62, pp. 305–308, 1971.

[197] B.E. Griffiths and K.A. Loparo, "Optimal control of jump linear Gaussian systems", *International Journal of Control* 42(4), pp. 791–819, 1985.

[198] D. Gross and D.R. Miller, "The randomization technique as a modelling tool and solution procedure for transient Markov processes", *Operations Research* 32(2), pp. 343–361, 1984.

[199] P.J. Haas and G.S. Shedler, "Regenerative generalized semi-Markov processes", *Commun. Statistics–Stochastic Models* 3(3), 1987.

[200] J.M. Hammersley and D.C. Handscomb, *Monte Carlo Methods*, Methuen, London, 1964.

[201] B.R. Haverkort, *Performability Modelling Tools, Evaluation Techniques, and Applications*, Ph.D. Thesis, Department of Computer Science, University of Twente, 1990.

[202] B.R. Haverkort, "Approximate performability and dependability modelling using generalized stochastic Petri nets", *Performance Evaluation* 18(1), pp. 61–78, 1993.

[203] B.R. Haverkort, "Performability modelling using DYQNTOOL+", *International Journal of Reliability, Quality and Safety Engineering*, pp. 383–404, 1995.

[204] B.R. Haverkort, "In search for probability mass: probabilistic evaluation of high-level specified Markov models", *Computer Journal* 38(7), pp. 521–529, 1995.

[205] B.R. Haverkort, "Matrix-geometric solution of infinite stochastic Petri nets", in *Proceedings of the 1st International Computer Performance and Dependability Symposium*, IEEE Computer Society Press, pp.72–81, 1995.

[206] B.R. Haverkort, "Approximate analysis of networks of PH|PH|1|K queues: theory & tool support", in *Quantitative Evaluation of Computing and Communication Systems*, H. Beilner, F. Bause, editors, *Lecture Notes in Computer Science* 977, Springer Verlag, pp. 239–253, 1995.

[207] B.R. Haverkort, "SPN2MGM: Tool support for matrix geometric stochastic Petri nets", *Proceedings of the 2nd International Computer Performance and Dependability Symposium*, IEEE Computer Society Press, pp. 219–228, 1996.

[208] B.R. Haverkort, *Performance of Computer-Communication Systems: a Model-Based Approach*, John Wiley & Sons, 1998.

[209] B.R. Haverkort, "Approximate analysis of networks of PH|PH|1|K queues with customer losses: test results", *Annals of Operations Research* 79, pp. 271–291, 1998.

[210] B.R. Haverkort, A. Bell and H.C. Bohnenkamp, "On the efficient sequential and distributed generation of very large Markov chains from stochastic Petri nets", *Proceedings of the 8th International Workshop on Petri Nets and Performance Models*, pp. 12–21, IEEE Computer Society Press, Zaragoza, Spain, 1999.

[211] B.R. Haverkort, H. Hermanns and J.-P. Katoen, "On the use of model checking techniques for dependability evaluation", in *Proceedings of the 19th Symposium on Reliable Distributed Systems*, IEEE CS Press, pp. 228–237, 2000.

[212] B.R. Haverkort and I.G. Niemegeers, "Using dynamic queueing networks for performability modelling", in *Proceedings of the European Simulation Multiconference*, Society for Computer Simulation, pp. 184–191, 1990.

[213] B.R. Haverkort and I.G. Niemegeers, "Performability modelling tools and techniques", *Performance Evaluation* 25(1), pp. 17–40,1996.

[214] B.R. Haverkort, I.G. Niemegeers and P. Veldhuyzen van Zanten, "DyQNtool: a performability modelling tool based on the dynamic queueing network concept", in *Proceedings of the Fifth International Conference on Computer Performance Evaluation: Modelling Techniques and Tools*, G. Balbo, G. Serazzi, editors, North-Holland, pp. 181–195, 1992.

[215] B.R. Haverkort and A.M.H. Meeuwissen, "Sensitivity and uncertainty analysis of Markov reward models", *IEEE Transactions on Reliability* 44(1), pp. 147–154, 1995.

[216] B.R. Haverkort, A.P.A. van Moorsel and D.-J. Speelman, "Xmgm: A performance analysis tool based on matrix geometric methods", in *Proceedings MASCOTS 1994*, IEEE Computer Society Press, pp. 152–157, 1994.

[217] B.R. Haverkort and K.S. Trivedi, "Specification techniques for Markov reward models", *Discrete Event Dynamic Systems: Theory and Application* 3, pp. 219–247, July 1993.

[218] P. Heidelberger, "Fast simulation of rare events in queueing and reliability models", *ACM Transactions on Modeling and Computer Simulation* 5(1), pp. 43–85, 1995.

[219] P. Heidelberger and A. Goyal, "Sensitivity analysis of continuous time Markov chains using uniformization", in *Computer Performance and Reliability*, G. Iazeolla, P.J. Courtois and O.J. Boxma, editors, North-Holland, pp. 93–104, 1988.

[220] P. Heidelberger, V.F. Nicola and P. Shahabuddin, "Simultaneous and efficient simulation of highly dependable systems with different underlying distributions", in *Proceedings of the 1992 Winter Simulation Conference*, IEEE Computer Society Press, pp. 458–465, 1992.

[221] P. Heidelberger, P. Shahabuddin and V.F. Nicola, "Bounded relative error in estimating transient measures in highly dependable non-Markovian systems", *ACM Transactions on Modeling and Computer Simulation* 4, pp. 137–164, 1994.

[222] A. Heindl and R. German, "A fourth-order algorithm with automatic stepsize control for the transient analysis of DSPNs", *Proceedings of the 7th International Workshop on Petri Nets and Performance Models*, pp. 60–69, IEEE Computer Society Press, 1997.

[223] H. Hermanns, *Interactive Markov Chains*, Ph.D. thesis, University of Erlangen-Nürnberg, 1999.

[224] H. Hermanns, U. Herzog, U. Klehmet, V. Mertsiotakis and M. Siegle, "Compositional performance modelling with the TIPPTOOL", *Performance Evaluation* 39, pp. 5–35, 2000.

[225] H. Hermanns, U. Herzog and J.-P. Katoen, "Process algebra for performance evaluation", *Theoretical Computer Science*, forthcoming.

[226] H. Hermanns and J.-P. Katoen, "Automated compositional Markov chain generation for a plain-old telephone system", *Science of Computer Programming* 36(1), pp. 97–127, 2000.

[227] H. Hermanns and M. Siegle, "Bisimulation algorithms for stochastic process algebras and their BDD-based implementation", in *Proceedings of ARTS'99*, J.-P. Katoen editor, *Lecture Notes in Computer Science* 1601, Springer Verlag, pp. 244–264, 1999.

[228] R. Hersh, "Random evolutions; a survey of results and problems", *Rocky Mountain J. Math.* 4, pp. 443–476, 1974.

[229] R. Hersh and M. Pinsky, "Random evolutions are asymptotically Gaussian", *Communications of Pure and Applied Mathematics* 25, pp. 33-44, 1972.

[230] J. Hillston, *A Compositional Approach to Performance Modelling*, Ph.D. thesis, University of Edinburgh, 1994.

[231] C. Hirel, S. Wells, R. Fricks and K.S. Trivedi, "iSPN: an integrated environemnt for modeling using stochastic Petri nets" in *Tools Descriptions of the 9th International Conference on Modelling Techniques and Tools for Computer Performance Evaluation*, pp. 17–19, Saint Malo, France, 1997.

[232] A.C.M. Hopmans and J.P.C. Kleijnen, "Importance sampling in system simulation: a practical failure?", *Mathematics and Computing in Simulation XXI*, pp. 209–220, 1979.

[233] A. Hordijk, D.L. Iglehart and R. Schassberger, "Discrete time methods for simulating continuous time Markov chains", *Advances in Applied Probability* 8, pp. 772–788, 1976.

[234] R.A. Horn and C.R. Johnson, *Topics in Matrix Analysis*, Cambridge University Press, 1991.

[235] G. Horton and S. Leutenegger, "A multi-level solution algorithm for steady-state Markov chains", *ACM Performance Evaluation Review* 22(1): 191–200, 1994.

[236] R.A. Howard, *Dynamic Probabilistic Systems, Vol. II: Semi-Markov and Decision Processes*, John Wiley & Sons, New York, 1971.

[237] R. Huslende, "A combined evaluation of performance and reliability for degradable systems", *ACM/SIGMETRICS Conference on Measurement and Modeling of Computer Systems*, pp. 157–164, ACM, 1981.

[238] O.C. Ibe, H. Choi and K.S. Trivedi, "Performance evaluation of client-server systems", *IEEE Transactions on Parallel and Distributed Systems* 4(11), pp. 1217–1229, 1993.

[239] D.L. Iglehart and G.S. Shedler, "Simulation of non-Markovian systems", *IBM Journal of Research and Development* 27(5), pp. 472–480, 1983.

[240] B.R. Iyer, L. Donatiello and P. Heidelberger, "Analysis of performability for stochastic models of fault-tolerant systems", *IEEE Transactions on Computers* 35(10), pp. 902–907, 1986.

[241] D.L. Jagerman, "An inversion technique for the Laplace transform", *Bell System Technical Journal* 61, pp. 1995–2002, October 1982.

[242] A. Jensen, "Markoff chains as an aid in the study of Markoff processes", *Skandinavsk Aktuarietidskrift* 36, pp. 87–91, 1953.

[243] A.M. Johnson Jr. and M. Malek, "Survey of software tools for evaluating reliability, availability, and serviceability", *ACM Computing Surveys* 20(4), pp. 227–269, 1988.

[244] G. Juanole and Y. Atamna, "Dealing with arbitrary time distributions with stochastic timed Petri net models – Application to queueing systems", in *Proceedings International Workshop on Petri Nets and Performance Models*, IEEE Computer Society Press, pp. 32–41, 1991.

[245] S. Juneja, *Efficient Rare Event Simulation of Stochastic Systems*, Ph.D. Thesis, Department of Operations Research, Stanford University, 1993.

[246] S. Juneja and P. Shahabuddin, "Fast simulation of Markovian reliability/availability models with general repair policies", *Proceedings of the Twenty-Second International Symposium on Fault-Tolerant Computing*, IEEE Computer Society Press, pp. 150–159, 1992.

[247] S. Juneja and P. Shahabuddin, "Efficient simulation of Markov chains with small transition probabilities", Research Report, Dept. of Industrial Engineering and Operations Research, Columbia University, 1997.

[248] S. Juneja and P. Shahabuddin, "A splitting based importance sampling algorithm for the fast simulation of Markov chains with small transition probabilities", Research Report, Dept. of Industrial Engineering and Operations Research, Columbia University, 1998.

[249] S. Juneja and P. Shahabuddin, "Simulating heavy tailed processes using delayed hazard rate twisting", Research Report, Dept. of Industrial Engineering and Operations Research, Columbia University, 1999.

[250] D. Kahner, C. Moler and S. Nash, *Numerical Methods and Software*, Prentice-Hall, 1989.

[251] H. Kahn and T.E. Harris, "Estimation of particle transmission by random sampling", *National Bureau of Standards Applied Mathematics Series* 12, pp. 27–30, 1951.

[252] H. Kahn and A.W. Marshall, "Methods of reducing sample size in Monte Carlo computations", *Journal of the Operations Research Society* 1(5), pp. 263–278, 1953.

[253] V.V. Kalashnikov, "Analytical and simulation estimates of reliability for regenerative models", *Syst. Anal. Model. Simul.* 6, pp. 833–851, 1989.

[254] K. Kant, *Introduction to Computer System Performance Evaluation*, McGraw-Hill, 1992.

[255] L.A. Kant and W.H. Sanders, "Loss process analysis of the knockout switch using stochastic activity networks", in *Proceedings of the 4th International Conference on Computer Communications and Networks*, Las Vegas, NV, pp. 344–349, Sept. 1995.

[256] S. Karlin and H.M. Taylor, *A First Course in Stochastic Processes*, Second Edition, Academic Press, New York, 1975.

[257] J. Keilson, *Markov Chain Models–Rarity and Exponentiality*, Springer Verlag, New York, 1979.

[258] J. Keilson and A. Kester, "Monotone matrices and monotone Markov processes", *Stochastic Processes and Their Applications* 5, pp. 231–245, 1977.

[259] J. Keilson and S.S. Rao, "A process with chain dependent growth rate", *Journal of Applied Probability* 7, pp. 699–711, 1970.

[260] J. Keilson and S.S. Rao, "A process with chain dependent growth rate. Part II: The ruin and ergodic problems", *Advances in Applied Probability* 3, pp. 315–338, 1971.

[261] Ch. Kelling, "TimeNET-SIM—a parallel simulator for stochastic Petri nets", in *Proceedings of the 28th Annual Simulation Symposium*, Society for Computer Simulation, Phoenix, AZ, pp. 250–258, 1995.

[262] P. Kemper, "Transient analysis of superposed GSPNs", in *Proceedings of the 7th International Workshop on Petri Nets and Performance Models*, IEEE Computer Society Press, pp. 101–110, 1997.

[263] G. Kesidis and J. Walrand, "Quick simulation of ATM buffers with on-off multiclass Markov fluid sources", *ACM Transactions on Modeling and Computer Simulation* 3, pp. 269–276, 1993.

[264] H.M. Khelalfa and A.K. von Mayrhauser, "Models to evaluate trade-offs between performance and reliability", in *Proceedings CMG XV*, San Francisco, pp. 24–29, 1984.

[265] G. Klas, *Hierarchical Evaluation of Generalized Stochastic Petri Nets*, Ph.D. thesis, Technical University München, 1993.

[266] W. Knottenbelt, M. Mestern, P. Harrison and P. Kritzinger, "Probability, parallelism and the state space exploration problem" in *Computer Performance Evaluation*, R. Puigjaner, N.N. Savino, and B. Serra, editors, *Lecture Notes in Computer Science* 1469, Springer Verlag, pp. 165–179, 1998.

[267] D.E. Knuth, "Big omicron, big omega and big theta", *ACM SIGACT News* 8(2), pp. 18–24, 1976.

[268] U. Krieger, B. Müller-Clostermann and M. Sczittnick, "Modelling and analysis of communication systems based on computational methods for Markov chains", *IEEE Journal on Selected Areas in Communications* 8(9), pp. 1630–1648, 1990.

[269] V.G. Kulkarni, *Modeling and Analysis of Stochastic Systems*, Chapman and Hall, 1995.

[270] V.G. Kulkarni, V.F. Nicola, R.M. Smith and K.S. Trivedi, "Numerical evaluation of performability and job completion time in repairable fault-tolerant systems", in *Proceedings of the Sixteenth International Symposium on Fault-Tolerant Computing*, IEEE Computer Society Press, pp. 252–257, 1986.

[271] V.G. Kulkarni, V.F. Nicola and K.S. Trivedi, "On modeling the performance and reliability of multi-mode computer systems", *Journal of Systems and Software* 6(1&2), pp. 175–183, 1986.

[272] V.G. Kulkarni, V.F. Nicola and K.S. Trivedi, "The completion time of a job on multimode systems", *Advances in Applied Probability* 19, pp. 932–954, 1987.

[273] J.-C. Laprie, editor, *Dependability: Basic Concepts and Terminology*, Volume 5 of *Dependable Computing and Fault-Tolerant Systems*, Springer Verlag, 1992.

[274] J.C. Laprie, "Dependable computing and fault-tolerance: concepts and terminolgy", in *Proceedings of the 15th International Fault-Tolerant Computing Symposium*, IEEE Computer Society Press, pp. 2–7, 1985.

[275] P. l'Ecuyer and Y. Champoux, "Importance sampling for large ATM type queuing networks", in *Proceedings of the 1996 Winter Simulation Conference*, IEEE Press, pp. 309–316, 1996.

[276] R.M.M. Leão, E. de Souza e Silva and S.C. de Lucena, "A set of tools for traffic modeling, analysis and experimentation", in *11th International Conference on Modelling Techniques and Tools for Computer Performance Evaluation*, B. Haverkort, H. Bohnenkamp and C. Smith, editors, *Lecture Notes in Computer Science* 1786, Springer Verlag, pp. 40–55, 2000.

[277] Y.H. Lee and K.G. Shin, "Optimal reconfiguration strategy for a degradable multimode computer system", *Journal of the ACM* 34(2), pp. 326–348, 1987.

[278] T. Lehtonen and H. Nyrhinen, "Simulating level-crossing probabilities by importance sampling", *Advances in Applied Probability* 24, pp. 858–874, 1992.

[279] R. Lepold, "TOMPSIN: Benutzerhandbuch", Siemens AG, 1991.

[280] R. Lepold, "PENPET: a performability modelling evaluation tool based on stochastic Petri nets", in *Proceedings of the Joint International Meeting of TIMS XXX and SOBRAPO XXIII*, Rio de Janeiro, 1991.

[281] R. Lepold, "Performability evaluation of fault-tolerant computer systems using stochastic Petri nets", in *Proceedings of the Fifth International Conference on Fault-Tolerant Computing Systems*, Nürnberg, 1991.

[282] R. Lepold, *Performability Evaluation of Degradable Computer Systems Based on Stochastic Petri Nets*, Ph.D. thesis, Université de Haute Alsace, France, 1992.

[283] Y. Levy and P.E. Wirth, "A unifying approach to performance and reliability objectives", in *Proceedings of the 12th International Teletraffic Congress*, pp. 4.2B2.1–4.2B2.7, Torino, 1988.

[284] E.E. Lewis and F. Bohm, "Monte Carlo simulation of Markov unreliability models", *Nuclear Engineering and Design* 77, pp. 49–62, 1984.

[285] P.A.W. Lewis and G.S. Shedler, "Simulation of non-homogeneous Poisson processes by thinning", *Naval Research Logistics Quarterly* 26(3), pp. 403–413, 1979.

[286] Y. Li, *Analysis of Markov Reward Models of Fault-Tolerant Computer Systems*, M.S. Thesis, Department of Electrical and Systems Engineering, University of Connecticut, Storrs, CT, 1990.

[287] C.A. Liceaga and D.P. Siewiorek, "Towards automatic Markov reliability modeling of computer architectures", *NASA Technical Memorandum* 89009, 1986.

[288] D. Lieber, R.Y. Rubinstein and D. Elmakis, "Quick estimation of rare events in stochastic networks", *IEEE Transactions on Reliability* 46(2), pp. 254–265, 1994.

[289] C. Lindemann, "DSPNEXPRESS: a software package for the efficient solution of deterministic and stochastic Petri nets", in *Computer Performance Evaluation '92, Modelling Techniques and Tools*, R. Pooley, J. Hillston, editors, Antony Rowe Ltd, Chippenham, UK, pp. 13–30, 1992.

[290] C. Lindemann, "An improved numerical algorithm for calculating steady-state solutions of deterministic and stochastic Petri net models", *Performance Evaluation* 18, pp. 75–95, 1993.

[291] C. Lindemann, "DSPNEXPRESS: a software package for the efficient solution of deterministic and stochastic Petri nets", *Performance Evaluation* 22, pp. 3–21, 1995.

[292] C. Lindemann, *Performance Modelling with Deterministic and Stochastic Petri Nets*, John Wiley & Sons, 1998.

[293] C. Lindemann and G.S. Shedler, "Numerical analysis of deterministic and stochastic Petri nets with concurrent deterministic transitions", *Performance Evaluation* 27&28, pp. 565–583, 1996.

[294] L.M. Malhis, W.H. Sanders and R.D. Schlichting, "Numerical evaluation of a group-oriented multicast protocol using stochastic activity networks", in *Proceedings of the 6th International Workshop on Petri Nets and Performance Models*, Durham, NC, pp. 63–72, 1995.

[295] L.M. Malhis, S.C. West, L.A. Kant and W.H. Sanders, "Modeling recycle: A case study in the industrial use of measurement and modeling", in *Proceedings International Computer Performance and Dependability Symposium*, Erlangen, Germany, pp. 285–294, 1995.

[296] R. Mallubhatla, K.R. Pattipati and N. Viswanadham, "Moment recursions of the cumulative performance of production systems using discrete-time Markov reward models", in *Proceedings of the 1994 IEEE International Conference on Robotics and Automation*, San Diego, CA, pp. 1812–1817, 1994.

[297] R. Mallubhatla and K.R. Pattipati, "Discrete-time Markov reward models: random rewards", in *Proceedings of the Rensselaer's Fourth International Conference on Computer Integrated Manufacturing and Automation Technology*, Troy, NY, pp. 315–330, June 1994.

[298] R. Mallubhatla, K.R. Pattipati and N. Viswanadham, "Discrete-time Markov reward models of production systems," *Discrete Event Systems, Manufacturing Systems and Communication Networks*, P.R. Kumar and P. Varaiya, editors, Springer Verlag, pp. 149–175, 1995.

[299] M. Mandjes and A. Ridder, "Finding the conjugate of Markov fluid processes", *Probability in the Engineering and Informational Sciences* 9, pp. 297–315, 1994.

[300] R.A. Marie, A.L. Reibman and K.S. Trivedi, "Transient analysis of acyclic Markov chains", *Performance Evaluation* 7, pp. 175–194, 1987.

[301] M. Mariton, "Jump linear quadratic control with random state discontinuities", *Automatica* 23(2), pp. 237–240, 1987.

[302] W.A. Massey, "Stochastic ordering for Markov processes on partially ordered spaces", *Math. Oper. Res.* 12, pp. 350–367, 1987.

[303] R.A. McLean and M.F. Neuts, "The integral of a step function defined on a Semi-Markov process", *SIAM Journal on Applied Mathematics* 15, pp. 726–737, 1967.

[304] H. de Meer, *Transiente Leistungsbewertung und Optimierung Rekonfigurierbarer Fehlertoleranter Rechensysteme*, Ph.D. Thesis, Friedrich-Alexander Universität Erlangen-Nürnberg, 1992.

[305] H. de Meer and O.-R. Düsterhöft, "Controlled stochastic Petri nets", in *Proceedings of the 16th Symposium on Reliable Distributed Systems*, IEEE Computer Society Press, pp. 18–25, 1997.

[306] H. de Meer and S. Fischer, "Controlled stochastic Petri nets for QoS management", in *Proceedings of the 9th ITG/GI Conference on Measurement, Modeling and Performance Evaluation of Computer and Communication Systems*, VDE Verlag, pp. 161–172, 1997.

[307] H. de Meer and H. Mauser, "A modeling approach for dynamically reconfigurable computer systems", in *Dependable Computing and Fault Tolerant Systems*, Springer Verlag, pp. 149–158, 1993.

[308] H. de Meer and H. Sevcikova, "PENELOPE: dependability evaluation and the optimization of performability", in *Computer Performance Evaluation: Modelling Techniques and Tools*, R. Marie, B. Plateau, M. Calzarossa, G. Rubino, editors, *Lecture Notes in Computer Science* 1245, Springer Verlag, pp. 19–31, 1997.

[309] H. de Meer, K.S. Trivedi and M. Dal Cin, "Guarded repair of dependable systems", *Theoretical Computer Science* 128, pp. 179–210, 1994.

[310] B. Melamed and M. Yadin, "Numerical computation of sojourn-time distributions in queueing networks", *Journal of the ACM* 31(4), pp. 839–854, 1984.

[311] B. Melamed and M. Yadin, "Randomization procedures in the computation of cumulative-time distributions over discrete state Markov processes", *Operations Research* 32(4), pp. 926–944, 1984.

[312] M. Meo, E. de Souza e Silva and M. Ajmone Marsan, "Efficient solution of a class of Markov chain models of telecommunication systems", *Performance Evaluation* 27/28, pp. 603–625, 1996.

[313] V. Mertsiotakis, *Approximate Analysis Methods for Stochastic Process Algebras*, Ph.D. Thesis, Friedrich-Alexander Universität Erlangen-Nürnberg, 1998.

[314] J.F. Meyer, "On evaluating the performability of degradable computing systems", in *Proceedings 8th International Symposium on Fault-Tolerant Computing*, Toulouse, France, pp. 44–49, June 1978.

[315] J.F. Meyer, "On evaluating the performability of degradable computer systems", *IEEE Transactions on Computers* 29(8), pp. 720–731, 1980.

[316] J.F. Meyer, "Closed-form solutions of performability", *IEEE Transactions on Computers* 31(7), pp. 648–657, 1982.

[317] J.F. Meyer, "Performability modeling of distributed real time systems", in *Mathematical Computer Performance and Reliability*, G. Iazeolla, P.J. Courtois, and A. Hordijk, editors, North-Holland, 1984.

[318] J.F. Meyer, A. Movaghar and W. H. Sanders, "Stochastic activity networks: Structure, behavior and application", in *Proceedings International Workshop on Timed Petri Nets*, IEEE Computer Society Press, pp. 106–115, 1985.

[319] J.F. Meyer, "Performability: a retrospective and some pointers to the future", *Performance Evaluation* 14(3&4), pp. 139–156, 1992.

[320] J.F. Meyer, "Performability evaluation: where it is and what lies ahead", *Proceedings of the IEEE International Performance and Dependability Symposium*, pp. 334–343, 1995.

[321] J.F. Meyer, K.H. Muralidhar and W.H. Sanders, "Performability of a token bus network under transient fault conditions", in *Proceedings 19th International Fault-Tolerant Computing Symposium*, Chicago, IL, pp. 175–182, June 1989.

[322] J.F. Meyer and L. Wei, "Influence of workload on error recovery in random access memories", *IEEE Transactions Computers* 37(4), pp. 500–507, 1988.

[323] I. Mitrani and B. Avi-Itzhak, "A many-server queue with service interruptions", *Operations Research* 16, pp. 628–638, 1968.

[324] I. Mitrani and R. Chakka, "Spectral expansion solution for a class of Markov models: Application and comparison with the matrix-geometric method", *Performance Evaluation* 23, pp. 241–260, 1995.

[325] I. Mitrani and D. Mitra, "A spectral expansion method for random walks on semi infinite strips", *IMACS Symposium on Iterative Methods in Linear Algebra*, Brussels, 1991.

[326] I. Mitrani and A. Puhalskii, "Limiting results for multiprocessor systems with breakdowns and repairs", *Queueing Systems* 14, pp. 293–311, 1993.

[327] I. Mitrani and P.E. Wright, "Routing in the presence of breakdowns", *Performance Evaluation* 20, pp. 151–164, 1994.

[328] M.K. Molloy, "Performance analysis using stochastic Petri nets", *IEEE Transactions on Computers* 31, pp. 913–917, 1982.

[329] A.P.A. van Moorsel, B.R. Haverkort and I.G. Niemegeers, "Fault injection simulation: a variance reduction technique for systems with rare events", *Dependable Computing for Critical Applications 2*, Springer Verlag, pp. 115–134, 1991.

[330] A.P.A. van Moorsel and B.R. Haverkort, "Probabilistic evaluation for the analytical solution of large Markov chains: Algorithms and tool support", *Microelectronics and Reliability* 6, pp. 733–755, 1996.

[331] A.P.A. van Moorsel and Y. Huang, "Reusable software components for performability tools and their utilization for web-based configurable tools", in: *Computer Performance Evaluation: Modelling Techniques and Tools*, R. Puigjaner, N.N. Savino, B. Serra, editors, *Lecture Notes in Computer Science* 1469, Springer Verlag, pp. 37–50, 1998.

[332] A.P.A. van Moorsel and W.H. Sanders, "Adaptive uniformization", *Communications in Statistics–Stochastic Models* 10(3), pp. 619–648, 1994.

[333] A. Movaghar and J.F. Meyer, "Performability modeling with stochastic activity networks", in *Proceedings of 1984 Real-Time Systems Symposium*, Austin, TX, pp. 215–224, Dec. 1984.

[334] A.P.A. van Moorsel, *Performability Evaluation Concepts and Techniques*, Ph.D. Thesis, University of Twente, Enschede, Netherlands, 1993.

[335] A.P.A. van Moorsel and W.H. Sanders, "Adaptative uniformization", *Memoranda Informatica* 92-88, University of Twente, Enschede, Netherlands, 1993.

[336] B. Müller-Clostermann, "NUMAS – A tool for the numerical analysis of computer systems", in *Proceedings of the International Conference on Modelling Techniques and Tools for Performance Analysis*, D. Potier, editor, North-Holland, pp. 141–154, 1985.

[337] M. Mulazzani and K.S. Trivedi, "Dependability prediction: comparison of tools and techniques", in *Proceedings IFAC Safecomp*, pp. 171–178, 1986.

[338] M. Mulazzani, "An open layered architecture for dependability analysis and its applications", in *Proceedings of the 18th Fault-Tolerant Computing Symposium*, IEEE Computer Society Press, 1988.

[339] R.R. Muntz, E. de Souza e Silva and A. Goyal, "Bounding availability of repairable computer systems", *IEEE Transactions on Computers* 38(12), pp. 1714–1723, 1989.

[340] J.K. Muppala and K.S. Trivedi, "Numerical transient solution of finite Markovian queueing systems", in *Queueing and Related Models*, U. N. Bhat and I. V. Basawa, editors, Oxford University Press, pp. 262–284, 1992.

[341] H. Nabli and B. Sericola, "Performability analysis of fault-tolerant computer systems", *Research Report* 2254, INRIA, France, May 1994.

[342] H. Nabli and B. Sericola, "Performability analysis: a new algorithm", *IEEE Transactions on Computers* 45, pp. 491–494, 1996.

[343] M.K. Nakayama, *Simulation of Highly Reliable Markovian and Non-Markovian Systems*, Ph.D. Thesis, Department of Operations Research, Stanford University, 1991.

[344] M.K. Nakayama, "Likelihood ratio derivative estimators in simulations of highly reliable Markovian systems", *Management Science* 41, pp. 524–554, 1995.

[345] M.K. Nakayama, "General conditions for bounded relative error in simulations of highly reliable Markovian systems", *Advances in Applied Probability* 28(3), pp. 687–727, 1993.

[346] M.K. Nakayama, "A characterization of the simple failure biasing method for simulations of highly reliable Markovian systems", *ACM Transactions on Modeling and Computer Simulation* 4(1), pp. 52–88, 1994.

[347] M.K. Nakayama, "Fast simulation methods for highly dependable systems", in *Proceedings of the 1994 Winter Simulation Conference*, IEEE Computer Society Press, pp. 221–228, 1994.

[348] M.K. Nakayama, A. Goyal and P.W. Glynn, "Likelihood ratio sensitivity analysis for Markovian models of highly dependable systems", *Operations Research* 42(1), pp. 137–157, 1994.

[349] S. Natkin, *Reseaux de Petri Stochastiques*, Ph.D. Thesis, CNAM-PARIS, June 1980.

[350] M.F. Neuts, *Matrix Geometric Solutions in Stochastic Models*, Johns Hopkins University Press, Baltimore, MD, 1981.

[351] M.F. Neuts and D.M. Lucantoni, "A Markovian queue with N servers subject to breakdowns and repairs", *Management Science* 25, pp. 849–861, 1979.

[352] M.F. Neuts, "Two further closure properties of PH-distributions", *Asia-Pacific Journal of Operational Research* 9, pp. 459–477, 1992.

[353] D.M. Nicol and P. Heidelberger, "Parallel simulation of Markovian queueing networks using adaptive uniformization", in *Proceedings 1993 ACM SIGMETRICS Conference*, pp. 135–145, 1993.

[354] V.F. Nicola, "A single server queue with mixed types of interruptions", *Acta Informatica* 23, pp. 465–486, 1986.

[355] V.F. Nicola, "Lumping in Markov reward processes", *IBM Research Report RC 14719*, Yorktown Heights, New York, 1989.

[356] V.F. Nicola, V.G. Kulkarni and K. Trivedi, "Queueing analysis of fault-tolerant computer systems", *IEEE Transactions on Software Engineering* 13, pp. 363–375, 1987.

[357] V.F. Nicola, A. Bobbio and K. Trivedi, "A unified performance reliability analysis of a system with a cumulative down time constraint", *Microelectronics and Reliability* 32, pp. 49–65, 1992.

[358] V.F. Nicola, M.K. Nakayama, P. Heidelberger and A. Goyal, "Fast simulation of dependability models with general failure, repair and maintenance processes", in *Proceedings of the 20th International Fault-Tolerant Computing Symposium*, IEEE Computer Society Press, pp. 491–498, 1990.

[359] V.F. Nicola, M.K. Nakayama, P. Heidelberger and A. Goyal, "Fast simulation of highly dependable systems with general failure and repair processes", *IEEE Transactions on Computers* 42(8), pp. 1440–1452, 1993.

[360] V.F. Nicola, P. Heidelberger and P. Shahabuddin, "Uniformization and exponential transformation: Techniques for fast simulation of highly dependable non-Markovian systems", in *Proceedings of the Twenty-Second International Symposium on Fault-Tolerant Computing*, IEEE Computer Society Press, pp. 130–139, 1992.

[361] V.F. Nicola, P. Shahabuddin, P. Heidelberger and P.W. Glynn, "Fast simulation of steady-state availability in non-Markovian highly dependable systems", in *Proceedings of the Twenty-Third International Symposium on Fault-Tolerant Computing*, IEEE Computer Society Press, pp. 38–47, 1993.

[362] V.F. Nicola, G.J. Hagesteijn and B.G. Kim, "Fast simulation of the Leaky Bucket algorithm", in *Proceedings of the 1994 Winter Simulation Conference*, IEEE Computer Society Press, pp. 266–273, 1994.

[363] V.F. Nicola and G.J. Hagesteijn, "Estimation of consecutive cell loss probability in ATM networks", *Third Workshop on Performance Modelling and Evaluation of ATM Networks*, Ilkley, West Yorkshire, UK, July 1995.

[364] V.F. Nicola, P. Shahabuddin and P. Nakayama, "Techniques for the fast simulation of models of highly dependable systems", Research Report, Dept. of Industrial Engineering and Operations Research, Columbia University, 1999.

[365] W.D. Obal and W.H. Sanders, "An environment for importance sampling based on stochastic activity networks", in *Proceedings of the 13th Symposium on Reliable Distributed Systems*, IEEE Computer Society Press, pp. 64–73, 1994.

[366] W.D. Obal II and W.H. Sanders, "Importance sampling in ULTRA-SAN", *Simulation* 62(2), pp. 98–111, 1994.

[367] W.D. Obal II and W.H. Sanders, "State-space support for path-based reward variables", *Performance Evaluation* 35, pp. 233–251, 1999.

[368] T.W. Page Jr., S.E. Berson, W.C. Cheng and R.R. Muntz, "An object-oriented modelling environment", *ACM Sigplan Notices* 24(10), pp. 287–296, 1989.

[369] S. Parekh and J. Walrand, "A quick simulation method for excessive backlogs in networks of queues", *IEEE Transactions on Automatic Control* 34(1) pp. 54–56, 1989.

[370] K.R. Pattipati and S.A. Shah, "On the computational aspects of performabilty models of fault-tolerant computer systems", *IEEE Transactions on Computers* 39, pp. 832–836, 1990.

[371] K.R. Pattipati, Y. Li and H.A.P. Blom, "A unified framework for the performability evaluation of fault-tolerant computer systems", *IEEE Transactions on Computers* 42(3), pp. 312–326, 1993.

[372] I. Petrovski, *Lectures on Partial Differential Equations*, 3rd edition, Sanders, Philadelphia, PA, 1967.

[373] M. Pinsky, "Random evolutions", in *Probabilistic Methods in Differential Equations, Lecture Notes in Mathematics* 451, Springer Verlag, pp. 88–89, 1975.

[374] M. Pinsky, *Lectures on Random Evolutions*, World Scientific, Singapore, 1991.

[375] K.H. Prodromides and W.H. Sanders, "Performability evaluation of CSMA/CD and CSMA/DCR protocols under transient fault conditions", *IEEE Transactions on Reliability* 42(1), pp. 116–127, 1993.

[376] A. Puliafito, O. Tomarchio and L. Vita, "Porting SHARPE on the web: Design and implementation of a network computing platform using JAVA", in *Computer Performance Evaluation: Modelling Techniques and Tools*, R. Marie, B. Plateau, M. Calzarossa, G. Rubino, editors, *Lecture Notes in Computer Science* 1245, Springer Verlag, pp. 32–43, 1997.

[377] P. S. Puri, "A method for studying the integral functionals of stochastic processes with applications: I. The Markov chain case", *Journal of Applied Probability* 8(2), pp. 331–343, 1971.

[378] M.A. Qureshi and W.H. Sanders, "Reward model solution methods with impulse and rate rewards: an algorithm and numerical results", *Performance Evaluation* 20(4), pp. 413–436, 1994.

[379] M.A. Qureshi and W.H. Sanders, "The effect of workload on the performance and availability of voting algorithms", *Microelectronics and Reliability* 36(6), pp. 757–774, 1996.

[380] M.A. Qureshi, W.H. Sanders, A.P.A. van Moorsel and R. German, "Algorithms for the generation of state-level representations of stochastic activity networks with general reward structures", in *Proceedings of the 6th International Workshop on Petri Nets and Performance Models*, Durham, NC, IEEE CS Press, pp. 180–190, 1995.

[381] K.G. Ramamurthy and N.K. Jaiswal, "A two-disimilar-unit cold standby system with allowed downtime", *Microelectronics and Reliability* 22, pp. 689–691, 1982.

[382] A.V. Ramesh and K.S. Trivedi, "Semi-numerical transient analysis of Markov models", in *Proceedings of the 33rd ACM Southeast Conference*, pp. 13–23, 1995.

[383] A. Reibman, "Modeling the effect of reliability on performance", *IEEE Transactions on Reliability* 39, pp. 314–320, 1990.

[384] A. Reibman, R. Smith and K.S. Trivedi, "Markov and Markov reward models transient analysis: an overview of numerical approaches", *European Journal of Operational Research* 40, pp. 257–267, 1989.

[385] A. Reibman and K.S. Trivedi, "Numerical transient analysis of Markov models", *Computers and Operations Research* 15(1), pp. 19–36, 1988.

[386] A. Reibman and K.S. Trivedi, "Transient analysis of cumulative measures of Markov model behavior", *Stochastic Models* 5(4), pp. 683–710, 1989.

[387] A. Reibman, K. Trivedi, S. Kumar and G. Ciardo, "Analysis of stiff Markov chains", *ORSA Journal on Computing* 1(2), pp. 126–133, 1989.

[388] S.M. Rezaul Islam and H.H. Ammar, "Performability of the hypercube", *IEEE Transactions on Reliability* 38(5), pp. 518-525, 1989.

[389] J.A. Rolia and K.C. Sevcik, "The method of layers", *IEEE Transactions on Software Engineering* 21(8), pp. 689–700, 1995.

[390] S.M. Ross, *Stochastic Processes*, John Wiley & Sons, New York, 1983.

[391] S.M. Ross and J. Schechtman, "On the first time a separately maintained parallel system has been down for a fixed time", *Naval Research Logistic Quarterly* 26, pp. 285–290, 1979.

[392] K.W. Ross, D.H.K Tsang and J. Wang, "Monte Carlo summation and integration applied to multiclass queueing networks", *Journal of the ACM* 41(6), pp. 1110–1135, 1994.

[393] B.L. Rozdestvenskii and N.N. Janenko, *System of Quasilinear Equations and Their Applications to Gas Dynamics*, American Mathematical Society, 1983.

[394] G. Rubino, "Network reliability evaluation", in *State-of-the-Art in Performance Modeling and Simulation*, K. Bagchi and J. Walrand, editors, Gordon-Breach Books, pp. 275–302, 1996.

[395] G. Rubino and B. Sericola, "Sojourn times in finite Markov processes", *Journal of Applied Probability* 27, pp. 744–756, 1992.

[396] G. Rubino and B. Sericola, "Interval availability analysis using operational periods", *Performance Evaluation* 14(3&4), pp. 257–272, 1992.

[397] G. Rubino and B. Sericola, "Interval availability distribution computation", in *Proceedings 23rd International Fault-Tolerant Computing Symposium*, pp. 48–55, 1993.

[398] J.S. Sadowsky, "Large deviations and efficient simulation of excessive backlogs in a GI/G/m queue", *IEEE Transactions on Automatic Control* 36(12), pp. 1383–1394, 1991.

[399] R.A. Sahner, *A Hybrid, Combinatorial-Markov Method for Solving Performance and Reliability Models*, Ph.D. Thesis, Duke University, 1986.

[400] R.A. Sahner and K.S. Trivedi, "Performance and reliability analysis using directed acyclic graphs", *IEEE Transactions on Software Engineering* 13(10), pp. 1105–1114, 1987.

[401] R.A. Sahner and K.S. Trivedi, "Reliability modelling using SHARPE", *IEEE Transactions on Reliability* 36(2), pp. 186–193, 1987.

[402] R.A. Sahner and K.S. Trivedi, "A hierarchical, combinatorial Markov method of solving complex reliability models", in *Proceedings of the IEEE-ACM Fall Joint Computer Conference*, IEEE Computer Society Press, pp. 817–825, 1986.

[403] R.A. Sahner and K.S. Trivedi, "A software tool for learning about stochastic models", *IEEE Transactions on Education* 36(1), pp. 56–61, 1993.

[404] R.A. Sahner, K.S. Trivedi and A. Puliafito, *Performance and Reliability Analysis of Computer Systems*, Kluwer Academic Press, 1996.

[405] W.H. Sanders, *Construction and Solution of Performability Models Based on Stochastic Activity Networks*, Ph.D. Thesis, University of Michigan, 1988.

[406] W.H. Sanders and R.S. Freire, "Efficient simulation of hierarchical stochastic activity network models", *Discrete Event Dynamic Systems: Theory and Application* 3(2&3), pp. 271–300, 1993.

[407] W.H. Sanders, L.A. Kant and A. Kudrimoti, "A modular method for evaluating the performance of picture archiving and communication systems", *Journal of Digital Imaging* 6(3), pp. 172–193, 1993.

[408] W.H. Sanders, R. Martinez, Y. Alsafadi and J. Nam, "Performance evaluation of a picture archiving and communication network using stochastic activity networks", *IEEE Transactions on Medical Imaging* 12(1), pp. 19–29, 1993.

[409] W.H. Sanders and J.F. Meyer, "METASAN: A performability evaluation tool based on stochastic activity networks", in *Proceedings of the ACM-IEEE Fall Joint Computer Conference*, IEEE Computer Society Press, pp. 807–816, 1986.

[410] W.H. Sanders and J.F. Meyer, "Performability evaluation of distributed systems using stochastic activity networks", in *Proceedings International Workshop on Petri Nets and Performance Models*, IEEE Computer Society Press, pp.111–120, 1987.

[411] W.H. Sanders and J.F. Meyer, "Performance variable driven construction methods for stochastic activity networks", in *Computer Performance and Reliability*, G. Iazeolla, P.-J. Courtois, O.J. Boxma, editors, North-Holland, pp.383–398, 1988.

[412] W.H. Sanders and J.F. Meyer, "A unified approach for specifying measures of performance, dependability, and performability", in *Dependable Computing for Critical Applications*, A. Avižienis and J.-C. Laprie, editors, Springer Verlag, pp. 215–237, 1991.

[413] W.H. Sanders and J.F. Meyer, "Reduced base model construction for stochastic activity networks", *IEEE Journal on Selected Areas in Communications* 9(1), pp. 25–36, 1991.

[414] W.H. Sanders, W.D. Obal II, M.A. Qureshi and F.K. Widjanarko, "ULTRASAN version 2: Architecture, features and implementation", *PMRL Technical Report* 93-17, University of Arizona, 1993.

[415] W.H. Sanders, W.D. Obal II, M.A. Qureshi and F.K. Widjanarko, "The *UltraSAN* modeling environment", *Performance Evaluation* 24, pp. 89–115, 1995.

[416] H. Schwetman, "CSIM: A C-based, process-oriented simulation language", in *Proceedings of the 1986 Winter Simulation Conference*, IEEE Press, pp. 387–396, 1986.

[417] H. Schwetman, "Using CSIM to model complex systems", in *Proceedings of the 1988 Winter Simulation Conference*, IEEE Press, pp. 491–499, 1988.

[418] M. Sczittnick and B. Müller-Clostermann, "Macom — A tool for the Markovian analysis of communication systems", in *Proceedings of the Fourth International Conference on Data Communication Systems and Their Performance*, pp. 456–470, 1990.

[419] B. Sengupta, "A queue with service interruptions in an alternating Markovian environment", *Operations Research* 38, pp. 308–318, 1990.

[420] B. Sericola, "Closed-form solution for the distribution of the total time spent in a subset of states of a homogeneous Markov process during a finite observation period", *Journal of Applied Probability* 27, pp. 713–719, 1990.

[421] B. Sericola, "Transient analysis of stochastic fluid models", Research Report 3152, INRIA, France, 1997.

[422] B. Sericola, "Availability analysis of repairable computer systems and stationarity detection", *IEEE Transactions on Computers* 48(11), pp. 1166–1172, 1999.

[423] P. Shahabuddin, *Simulation and Analysis of Highly Reliable Systems*, Ph.D. Thesis, Department of Operations Research, Stanford University, 1990.

[424] P. Shahabuddin, "Importance sampling for the simulation of highly reliable Markovian systems", *Management Science* 40, pp. 333–352, 1994.

[425] P. Shahabuddin, "Fast transient simulation of Markovian models of highly dependable systems", *Performance Evaluation* 20, pp. 267–286, 1994.

[426] P. Shahabuddin, "Rare event simulation", in *Proceedings of the 1995 Winter Simulation Conference*, IEEE Press, pp. 178–185, 1995.

[427] P. Shahabuddin and M.K. Nakayama, "Estimation of reliability and its derivatives for large time horizons in Markovian systems", *Proceedings of the 1993 Winter Simulation Conference*, IEEE Press, pp. 422–429, 1995.

[428] P. Shahabuddin and M.K. Nakayama, "Fast simulation techniques for estimating the unreliability in large regenerative models of highly reliable systems", Research Report, Department of Industrial Engineering and Operations Research, Columbia University, 1998.

[429] P. Shahabuddin, V.F. Nicola, P. Heidelberger, A. Goyal and P.W. Glynn, "Variance reduction in mean time to failure simulations", in *Proceedings of the 1988 Winter Simulation Conference*, IEEE Press, pp. 491–499, 1988.

[430] G. Shanthikumar, "First failure time of dependent parallel systems with safety period", *Microelectronics and Reliability* 26, pp. 955–972, 1986.

[431] J.G. Shanthikumar, "Uniformization and hybrid simulation/analytic models of reward processes", *Operations Research* 34(4), pp. 573–580, 1986.

[432] J.G. Shanthikumar and D.D. Yao, "General queueing networks: Representation and stochasticity", in *Proceedings of the 26th IEEE Conference on Decision and Control*, pp. 1084–1087, 1987.

[433] J.G. Shanthikumar and D.D. Yao, "Monotonicity properties in cyclic queueing networks with finite buffers", in *Proceedings of the 1st International Workshop on Queueing Networks with Blocking*, North Carolina, May 1988.

[434] J.G. Shanthikumar and D.D. Yao, "Throughput bounds for closed queueing networks with queue-independent service rates", *Performance Evaluation* 9, pp. 69–78, 1988.

[435] B.C. Shultes, *Regenerative Techniques for Estimating Performance Measures of Highly Dependable Systems with Repairs*, Ph.D. Thesis, School of Industrial and Systems Engineering, Georgia Institute of Technology, Atlanta, 1997.

[436] D. Siegmund, "Importance sampling in the Monte Carlo study of sequential tests", *The Annals of Statistics* 4, pp. 673–684, 1976.

[437] D. Siegmund, *Sequential Analysis: Tests and Confidence Intervals*, Springer Verlag, 1985.

[438] E. Smeitink, N.M. van Dijk and B.R. Haverkort, "Product forms for availability models", *Applied Stochastic Models and Data Analysis* 8, pp. 283–291, 1992.

[439] R.M. Smith and K.S. Trivedi, "The analysis of computer systems using Markov reward models," in *Stochastic Models of Computer and Communication Systems*, H. Takagi, editor, North-Holland, 1989.

[440] R.M. Smith, K.S. Trivedi and A.V. Ramesh, "Performability analysis: Measures, an algorithm and a case study", *IEEE Transactions on Computers* 37(4), pp. 406–417, 1988.

[441] A.D. Solovyev, "Asymptotic behaviour of the time of first occurrence of a rare event", *Engineering Cybernetics* 9, pp. 1038–1048, 1971.

[442] G.W. Stewart, "Computable error bounds for aggregated Markov chain", *Journal of the ACM* 30(2), pp. 271–285, 1983.

[443] W.J. Stewart, "MARCA: Markov chain analyzer: a software package for Markov modelling", in *Numerical Solution of Markov Chains*, W.J. Stewart, editor, Marcel Dekker, New York, pp. 37–62, 1991.

[444] W.J. Stewart, *Introduction to the Numerical Solution of Markov Chains*, Princeton University Press, NJ, 1994.

[445] W.J. Stewart and A. Goyal, "Matrix methods in large dependability models", IBM Research Report RC 11485, 1985.

[446] D. Stoyan, *Comparison Method for Queues and Other Stochastic Models*, John Wiley & Sons, 1983.

[447] S.G. Strickland, "Optimal importance sampling for quick simulation of highly reliable Markovian systems", in *Proceedings of the 1993 Winter Simulation Conference*, IEEE Press, pp. 437–444, 1993.

[448] S.G. Strickland, "Necessary and sufficient conditions for bounded relative error in importance sampling", University of Virginia, Charlottesville, Virginia, 1995.

[449] J. Stiffler and L. Bryant, "CARE III Phase III Report – Mathematical Description", NASA Contractor Report 3566, 1982.

[450] U. Sumita, J.G. Shantikumar and Y. Masuda, "Analysis of fault-tolerant computer systems", *Microelectronics and Reliability* 27, pp. 65–78, 1987.

[451] R. Suri, "A concept of monotonicity and its characterization for closed queueing networks", *Operations Research* 33, pp. 606–624, 1985.

[452] D.D. Sworder, "Control of systems subject to sudden change in character", *Proceedings of the IEEE* 64, pp. 1219–1225, 1976.

[453] A.T. Tai, J.F. Meyer and A. Avižienis, *Software Performability: From Concepts to Applications*, Kluwer, 1996.

[454] J.C. Tannenhill, "Hyperbolic and hyperbolic parabolic systems," Chapter 12 in *Handbook of Numerical Heat Transfer*, W.J. Minkowysz, E.M. Sparrow, E.E. Schneider and R.H. Fletcher, editors, Wiley, 1988, pp. 463–518.

[455] H. Tardif, K.S. Trivedi and A.V. Ramesh, "Closed-form transient analysis of Markov chains", Technical Report, Department of Computer Science, Duke University, Durham, NC, 1988.

[456] M. Telek and A. Bobbio, "Markov regenerative stochastic Petri nets with age type general transitions", in *Application and Theory of Petri Nets*, G. De Michelis and M. Diaz, editors, *Lecture Notes in Computer Science* 935, Springer Verlag, pp. 471–489, 1995.

[457] M. Telek, A. Bobbio, L. Jereb, A. Puliafito and K. Trivedi, "Steady state analysis of Markov regenerative SPN with age memory policy", in *Proceedings 8th International Conference on Modelling Techniques and Tools for Computer Performance Evaluation*, H. Beilner and F. Bause, editors, *Lecture Notes in Computer Science* 977, Springer Verlag, pp. 165–179, 1995.

[458] M. Telek, *Some Advanced Reliability Modelling Techniques*, Ph.D. Thesis, Hungarian Academy of Science, 1994.

[459] K. Thiruvengadam, "Queueing with breakdowns", *Operations Research* 11, pp. 62–71, 1963.

[460] H.C. Tijms, *Stochastic Modelling and Analysis, a Computational Approach*, Wiley, 1986.

[461] A.A. Törn, "Simulation nets, a simulation modeling and validation tool", *Simulation* 45(2), pp. 71–75, 1985.

[462] K.S. Trivedi, *Probability and Statistics with Reliability, Queueing and Computer Science Applications*, Prentice Hall, 1982.

[463] K.S. Trivedi, J.K. Muppala, S.P. Woolet and B.R. Haverkort, "Composite performance and dependability analysis", *Performance Evaluation* 14(3&4), pp.197–215, 1992.

[464] K. Trivedi, A.S. Sathaye, O.C. Ibe and R.C. Howe, "Should I add a processor?", in *Proceedings of the 23rd Annual Hawaii International Conference on System Sciences*, pp. 214–221, 1990.

[465] K. Trivedi, A. Reibman and R. Smith, "Transient analysis of Markov and Markov rewards models," in *Computer Performance and Reliability*, P.-J. Courtois, G. Iazeolla and O. Boxma, editors, North-Holland, pp. 535–545, 1988.

[466] P. Tsoucas and J. Walrand, "Monotonicity of throughput in non-Markovian networks", *Journal of Applied Probability* 26, pp. 134–141, 1989.

[467] M. Villen-Altamirano and J. Villen-Altamirano, "RESTART: a method for accelerating rare event simulation", in *Proceedings of the 13th International Teletraffic Congress*, North-Holland, pp. 71–76, 1991.

[468] M. Villen-Altamirano and J. Villen-Altamirano, "RESTART: a straightforward method for fast simulation of rare events", *Proceedings of the 1994 Winter Simulation Conference*, IEEE Press, pp. 282–289, 1994.

[469] M. Villen-Altamirano, A. Martinez Marron, J. Gamo and F. Fernandez-Cuesta, "Enhancement of accelerated simulation method RESTART by considering multiple threshholds", in *Proceedings of the 14th International Teletraffic Congress*, North-Holland, pp. 787–810, 1994.

[470] N. Viswanadham, Y. Narahari and R. Ram, "Performability of automated manufacturing systems", *Control and Dynamic Systems* 47, pp. 77–120, 1991.

[471] N. Viswanadham, K.R. Pattipati and V. Gopalakrishna, "Performability studies of AMSs with multiple part types", invited paper at the *1993 IEEE Robotics and Automation Conference*, Atlanta, GA, 1993.

[472] N. Viswanadham, K.R. Pattipati and V. Gopalakrishna, "Performability studies of automated manufacturing systems with multiple part types", *IEEE Transactions on Robotics and Automation* 11(5), pp. 692–709, 1995.

[473] R.F. Warming, P. Kutler and H. Lomax, "Second and third order noncentered difference schemes for nonlinear hyperbolic equations", *AIAA Journal* 11, pp. 189-196, 1973.

[474] H. Weisberg, "The distribution of linear combinations of order statistics from the uniform distribution", *Annals Math. Stat.* 42, pp. 704–709, 1971.

[475] C.J. Wenk and Y. Bar-Shalom, "A multiple model adaptive dual control algorithm for stochastic system with unknown parameters", *IEEE Transactions in Automatic Control* 25(4), pp. 703–710, 1980.

[476] W. Whitt, "Comparing counting processes and queues", *Advances in Applied Probability* 13, pp. 207–220, 1981.

[477] W. Whitt, "Stochastic comparison for non-Markov processes", *Math. Operations Research* 11(4), pp. 608–618, 1986.

[478] H.C. White and L.S. Christie, "Queueing with preemptive priorities or with breakdown," *Operations Research* 6, pp. 79–95, 1958.

[479] A. Zimmermann and J. Freiheit, "TimeNET-MS—An integrated modeling and performance evaluation tool for manufacturing systems", in *Proceedings of the IEEE International Conference on Systems, Man, and Cybernetics*, pp. 535–540, 1998.

[480] R. Zijal, *Analysis of Discrete-Time Deterministic and Stochastic Petri Nets*, Ph.D. Thesis, Technical University Berlin, 1997.

Concise glossary

ACE	acyclic Markov chain evaluator
CDF	cumulative density function
CP	cumulative performability
CTMC	continuous-time Markov chain
DSPN	discrete and stochastic Petri net
DTMC	discrete-time markov chain
EMRM	extended markov reward model
FFT	fast Fourier transform
FIFO	first-in, first-out
GMTF	general modelling tool framework
GSPN	generalized stochastic Petri net
LST	Laplace-Stieltjes transform
LU	lower-upper (decomposition)
MMAP	Markov-modulated arrival process
MRM	Markov reward model
MRSPN	Markov regenerative stochastic Petri net
MRTA	mean reward to absorption
MTBF	mean time between failures
PDE	partial differential equation
PDF	probability density function
PH	phase-type (distribution)
SAN	stochastic activity network
SOR	successive over-relaxation
SPA	stochastic process algebra
SPN	stochastic Petri net
SRN	stochastic reward nets
SSP	steady-state performability
TP	transient performability

Index

Printed and bound by CPI Group (UK) Ltd, Croydon, CR0 4YY

27/10/2024

14580294-0001